SpringerBriefs in Optimization

For further volumes:
www.springer.com/series/8918

Joseph Geunes

Demand Flexibility in Supply Chain Planning

Springer

Joseph Geunes
Industrial and Systems Engineering
University of Florida
Gainesville, FL
USA

ISSN 2190-8354 e-ISSN 2191-575X
SpringerBriefs in Optimization
ISBN 978-1-4419-9346-5 e-ISBN 978-1-4419-9347-2
DOI 10.1007/978-1-4419-9347-2
Springer New York Dordrecht Heidelberg London

Library of Congress Control Number: 2012933121

Mathematics Subject Classification (2010): 60C05, 60A10, 90B15, 90B99

Printed on acid-free paper

Springer is part of Springer Science+Business Media (www.springer.com)

For Paul & Vicki Geunes
and Sherry, Eric, and Brett

Preface

The goal of this book is to codify ideas and results from a specific segment of the supply chain operations planning literature that has developed over the past few decades. This segment of the literature draws on the tools of operations research in order to characterize optimal solutions to problems that seek to efficiently match a producer's supply output with the demands or requirements of a set of customers and/or markets. More specifically, we will emphasize contexts in which the producer has some control over both supply and demand, i.e., situations in which some degree of flexibility in demand exists from the producer's point of view.

The evolution of the operations literature in the past half century has by and large focused on managing (or minimizing) costs while attempting to meet some external party's target output requirements. This external party often corresponds to a marketing group within the same firm, whose responsibilities include setting prices and estimating the resulting customer demand levels, in effect, determining optimal demand levels with respect to some objective. Understanding how price influences demand for a good, and thus, what constitutes an *optimal* set of demand levels, requires some knowledge of how customers will respond to one of the product's critical characteristics (in this case, price). Defining the way in which customers will respond to price in the aggregate is analogous to characterizing the degree of flexibility that exists in demand as a function of price. Customer flexibility often exists along numerous product dimensions in addition to price (e.g., product sizes, delivery quantities, delivery lead times), many of which are directly controllable via production and distribution operations.

In addition to inherent customer flexibility with respect to product characteristics, a supplier or producer often has discretion as to which customers, demands, or markets it will satisfy with its product(s). This discretion provides an additional source of flexibility in planning by permitting the producer to accept or decline certain customers or markets.

Models for operations planning have typically treated demands as fixed, exogenous parameters, based on predetermined price levels and other fixed product characteristics. This corresponds to a sequential decision making process in which different decisions that ultimately combine to determine profitability are made separately. That is, marketing and sales groups essentially estimate the demand levels for products containing specific characteristics offered at specific prices, and operations is tasked with meeting the implied demands at the lowest delivered cost. An alternative view, which serves as the focus of this book, treats demand (and/or revenue) as dependent on key product-characteristic and customer-acceptance decisions that are made by the producer. This leads to new classes of operations planning models that effectively treat demand levels as decision variables within the planning model. The resulting models then determine the optimal production and demand levels, i.e., the most efficient match between the supply process and the inherently flexible demands. This book thus brings together several operations research based planning models that share this alternative view of sales and operations planning. As the final part of the book indicates, the models presented provide a foundation for both adapting a wealth of existing problems to this paradigm and for its extension and generalization to even broader classes of decision problems.

Gainesville, FL, USA Joseph Geunes

Acknowledgements

The majority of the contents of this book are the product of collaborative research over the past decade with numerous colleagues and collaborators who appear in the various references. Many of the key contributions and ideas presented throughout the book are attributable to these collaborators, and this book is an attempt to distill these contributions into a manageable and digestible form. I am forever indebted to my teachers, colleagues, and students, without whom this book and the research on which it is based could not exist. I would first like to acknowledge and thank Anant Balakrishnan, who instilled research standards which, despite eluding my grasp, serve as a goal to which I continue to strive. My primary collaborator (and friend) over the past decade has been Edwin Romeijn, who has taught me as much as anyone I have encountered. I only hope he has found our collaboration to be half as beneficial as have I. I would like to acknowledge the debt I owe to my department chairs Don Hearn and Joe Hartman for the support and freedom they have provided. I would also like to acknowledge several of the key contributors of important developments contained within this work, in particular, Wilco van den Heuvel, Wei Huang, Erhun Kundakcioglu, Retsef Levi, Tom Sharkey, Max Shen, David Shmoys, and Albert Wagelmans. In addition to these colleagues, I was fortunate to work with several insightful and energetic students on the problems contained in this book, including Semra Ağralı, Ismail Bakal, Yasemin Merzifonluoğlu, Chase Rainwater, and Kevin Taaffe. Finally, I would like to acknowledge the support provided by the National Science Foundation (grants #DMI-0322715 and #CMMI-0927930).

Contents

Part I
Supply Chain Operations Models with Demand Shaping

Chapter 1
Scope of Problem Coverage and Introduction

Abstract This chapter begins with an introduction to the book's scope and preliminary concepts applied throughout the book. We then present a set of basic, foundational, and classical models from the operations planning literature that serve as the underpinning of the work presented throughout the book. These models include the economic order quantity (EOQ), the newsvendor problem, the economic lot-sizing problem (ELSP), the knapsack problem (KP), the generalized assignment problem (GAP), and the facility location problem (FLP). The main results presented later in this book generalize these classical models to account for a planner's ability to influence demands, which have traditionally served as fixed parameters in these foundational models.

1.1 Scope and Preliminaries

The work in this book generalizes several of the most fundamental and classical models for production and inventory planning. These include the economic order quantity (EOQ), the newsvendor problem, the economic lot-sizing problem (ELSP), the knapsack problem (KP), the generalized assignment problem (GAP), and the facility location problem (FLP). Each of these models involves a very specific set of assumptions, which we will specify when introducing the associated model. Each model represents an abstraction with respect to some practical problem that is broadly applicable to entities that produce and/or stock consumer goods. This abstraction results in an idealized version of the associated real-world problem and, thus, one is unlikely to find that the required assumptions hold precisely in any practical setting. Despite this, these models are powerful for their approximation of reality and because they mathematically formalize important relationships among the key parameters and decision factors that combine to determine the economic performance of the system being modeled. This book will, therefore, present each model and its assumptions without providing strenuous arguments as to the degree to which these assumptions provide an effective approximation for any particular practical setting.

For ease of exposition, this book will also focus on the single-product version of the models in question. While multiple-product generalizations often follow based on a straightforward analysis, these generalizations tend to detract from the central

J. Geunes, *Demand Flexibility in Supply Chain Planning*,
SpringerBriefs in Optimization,
DOI 10.1007/978-1-4419-9347-2_1,

theme of modeling sources of demand flexibility. For the single-product version of a problem, we will essentially draw on three techniques to model a producer's ability to *shape* demand. The first of these techniques involves the ability to explicitly select a subset from some set of potential demands. We will refer to this technique as *demand selection*. We will refer to cases in which demand selection requires selecting all or none of a time-phased vector of demands as *market selection*. The second technique implicitly selects demands as a result of the dependence of demand on price. In this approach, selecting a demand level is equivalent to selecting a price level, assuming a one-to-one correspondence between price and demand in any planning period. The third technique we will explore may be characterized as a form of *demand sizing*. This method permits selecting the level, quantity, or size at which each demand will be satisfied, within some prespecified upper and lower limits. Observe that this concept of demand sizing generalizes demand selection when the prespecified lower limit equals zero. That is, choosing a size of zero under the demand sizing technique corresponds to a decision to not select a given demand, whereas choosing a positive size corresponds to selecting the demand for satisfaction.

1.2 Overview of Foundational Models

This section describes the basic models that serve as a basis for our exploration of operations models that incorporate demand flexibility. Each of the following six subsections provides a brief definition of a model that will be generalized in a later chapter.

1.2.1 The Economic Order Quantity (EOQ) Model

The economic order quantity (EOQ) model serves as the oldest quantitative model for production and inventory planning [8]. Despite its simplicity and high degree of abstraction, it remains widely used today, as it elegantly captures perhaps the most critical tradeoff inherent in inventory planning contexts between fixed order costs and inventory holding costs. We next provide an overview of the EOQ model assumptions and main results. For an in-depth derivation and analysis of the EOQ model, please see [9].

The EOQ model considers a single stage of inventory that stocks a single product with a constant and continuous demand rate of D units per unit time that will persist infinitely far into the future. The planner wishes to stock the item in order to ensure that all demands are met from stock as they occur. This is possible because the demand rate is deterministic and the stage replenishes from a supply source with a known and fixed (and finite) delivery lead time and with no capacity limit on the amount it can supply. Any time the stage orders a quantity of Q units from the supply source, all Q units are delivered after the fixed lead time. The planner pays C

dollars for each unit ordered and also pays a fixed order cost of S dollars each time a replenishment order is placed. The planner also accrues a holding cost for each unit held in inventory of H dollars per unit per unit time. The planner wishes to minimize the average cost per unit time over the infinite horizon while meeting all demands on time. It is straightforward to show that because all costs are time invariant, as is the demand rate, the planner's optimal policy requires periodically ordering batches of constant size (Q), and timing these replenishment orders to arrive precisely at the point in time at which the current on-hand inventory will reach zero. The average cost per unit time as a function of the order quantity Q, which we denote by $AC(Q)$, can be written as

$$AC(Q) = CD + \frac{SD}{Q} + H\frac{Q}{2}. \tag{1.1}$$

The first term in (1.1) captures the average variable purchase cost per unit time, while the second and third terms correspond to the average fixed order cost per unit time and the average holding cost per unit time, respectively. It is straightforward to show that $AC(Q)$ is strictly convex in Q for all $Q > 0$, which implies that the following stationary point serves as a strict global minimum for (1.1) among all positive Q values:

$$Q^* = \sqrt{\frac{2SD}{H}}. \tag{1.2}$$

Equation (1.2) is referred to as the economic order quantity and it captures the critical tradeoff between fixed order costs and holding costs. A high relative value of the fixed order cost S leads to a large batch size, which increases the time between orders. Conversely, a high relative value of the holding cost H reduces the batch size, which leads to a lower average inventory level. It is more than an interesting mathematical curiosity that, at the optimal (EOQ) batch size, the average setup cost per unit time is exactly matched to the average holding cost per unit time, i.e.,

$$\frac{SD}{Q^*} = H\frac{Q^*}{2} = \sqrt{\frac{SDH}{2}}. \tag{1.3}$$

This result has motivated numerous heuristic solution approaches for more complex inventory planning problems in an attempt to match average fixed order costs and holding costs per unit time as closely as possible (see [9]). As a result of (1.3), the value of $AC(Q)$ at the EOQ can be written compactly as

$$AC(Q^*) = CD + \sqrt{2SDH}. \tag{1.4}$$

The above equation (1.4) will come into play again in Chap. 3 when we consider EOQ-type models with demand selection.

Before concluding this section, we note that the above equations can easily be generalized to account for settings in which the batch of size Q is not delivered all at one instant following a fixed lead time, but is instead accumulated at a finite rate. In particular, if inventory is accumulated at a rate of P units per unit time (where we must assume that $P \geq D$ in order to be able to keep up with demand), then we can simply replace each instance of the parameter H in Eqs. (1.1)–(1.4) with $H' = 1 - D/P$ and all of the results we have discussed remain valid. The resulting model is often referred to as the EOQ problem with a finite production rate, or simply as the Economic Production Quantity (EPQ) model. While the EOQ model tends to be more appropriate for an inventory stage that orders in batches from an external supplier, the EPQ model tends to apply more readily to internal production environments when production contributes to inventory at a finite rate.

1.2.2 The Newsvendor Problem

The newsvendor problem is perhaps the simplest stochastic model for inventory planning. This problem considers a single planning period with uncertain demand. Demand in the period is therefore a random variable, which we denote as x_D, with an associated probability density function (pdf) of $f(x_D)$ and cumulative distribution function (cdf) $F(x_D)$ (our approach will assume that demand in the single period can be effectively modeled using a continuous probability distribution with mean μ_D and standard deviation σ_D). The planner stocks Q units in anticipation of the period's demand, where each unit comes at a procurement cost of C per unit. Stock is sold at a fixed unit price p. Any remaining stock at the end of the period is assessed a cost of H per unit, where a negative value of H corresponds to a salvage value and a positive value may be viewed as a disposal cost or, more generically, a holding cost. If demand exceeds the stock level Q, each shortage is assessed a penalty cost of B per unit (this "cost" may contain any lost profit margin in addition to any so-called loss-of-goodwill cost). The expected single-period profit, which depends on the quantity stocked, and which we denote by $\Pi(Q)$, is then written as

$$\Pi(Q) = p\left(\mu_D - \int_Q^{\infty} (x_D - Q) f(x_D)\,dx_D\right) - CQ$$
$$- H \int_0^Q (Q - x_D) f(x_D)\,dx_D - B \int_Q^{\infty} (x_D - Q) f(x_D)\,dx_D. \quad (1.5)$$

The first term on the right-hand side captures the expected revenue, where we have $\mathrm{E}[\mathrm{Sales}(Q)] = \mu_D - \int_Q^{\infty} (x_D - Q) f(x_D)\,dx_D$, and where $\mathrm{E}[\cdot]$ is the expected value operator. The second term denotes the variable purchase cost CQ, while the third and fourth terms capture the expected cost of leftovers and shortages, respectively, with $\mathrm{E}[\mathrm{Leftovers}(Q)] = \int_0^Q (Q - x_D) f(x_D)\,dx_D$ and $\mathrm{E}[\mathrm{Shortages}(Q)] = \int_Q^{\infty} (x_D - Q) f(x_D)\,dx_D$. This expected single-period profit equation can be equivalently written as

$$SPC(Q) = (p + H)\mu_D - (C + H)Q - (p + B + H) \int_Q^{\infty} (x_D - Q) f(x_D)\,dx_D. \quad (1.6)$$

Using Leibniz' rule [2], it is straightforward to show that $SPC(Q)$ is concave in Q and thus that the following stationary point provides its global maximum:

$$F(Q^*) = \frac{p - C + B}{p + B + H}. \tag{1.7}$$

We next assume that x_D is normally distributed with expected value μ_D and standard deviation σ_D (while the normal distribution has a support of $(-\infty, \infty)$, certain demand distributions may be effectively approximated using a normal distribution, assuming that the probability of negative demand is negligible; for example, if $3 \times \sigma_D < \mu_D$, then the probability of negative demand is less than 0.00135). Under normally distributed demand, (1.7) can be equivalently written as $\Phi(z^*) = (p - C + B)/(p + B + H)$, where $\Phi(z^*)$ is the cdf of the standard unit normal distribution at the critical fraction $(p - C + B)/(p + B + H)$, and z^* is the standard unit normal variate value at this fraction. We may then use the following equation to express the optimal order quantity:

$$Q^* = \mu_D + z^* \sigma_D. \tag{1.8}$$

The normal distribution assumption also allows us to rewrite the integral in (1.6) as

$$\int_Q^\infty (x_D - Q) f(x_D)\, dx_D = \sigma_D \int_z^\infty (u - z)\phi(u)\, du \equiv \sigma_D L(z), \tag{1.9}$$

where $z = (Q - \mu_D)/\sigma_D$ and $L(z)$ is known as the standard normal loss function (see [9]). Using (1.8) and (1.9), under normally distributed demand, we can write the expected profit, $\Pi_n(Q)$, at the optimal order quantity, Q^*, compactly as

$$\Pi_n(Q^*) = r\mu_D - K(z^*)\sigma_D, \tag{1.10}$$

where $r = p - C$ denotes the unit net revenue (profit margin) and $K(z^*) = (C + H)z^* + (p + B + H)L(z^*)$. Equation (1.10) will play a key role in our analysis of stochastic demand models with demand selection in Chap. 4. For one of the earliest papers dealing with the single-period inventory problem, please see [7].

1.2.3 The Economic Lot Sizing Problem (ELSP)

The economic lot sizing problem (ELSP) considers the same tradeoff as the EOQ model, using a discrete time approach with a finite number of time periods. This model assumes a finite horizon length of T periods, where D_t denotes the demand in period t, for $t = 1, \ldots, T$. Thus, demand is permitted to vary over time, although we are relegated to a discrete set of time points at which we may assess costs, in contrast with the EOQ model, which permits costs to accrue continuously throughout time. This discrete-time approach also allows handling time-varying cost parameters much more easily than with a continuous-time model. We denote S_t and

C_t, respectively, as the fixed order cost and the unit procurement cost in period t, for $t = 1, \ldots, T$. We must choose a convention for assessing inventory costs at some discrete set of time points; the most common convention in the literature applies a cost of H_t dollars per unit of inventory remaining at the end of period t, for $t = 1, \ldots, T$. In order to track costs, it is convenient to define Q_t and I_t as the order quantity and ending inventory in period t, for $t = 1, \ldots, T$. Assuming no demand is lost (i.e., all demand is met by the end of the horizon), we can represent inventory level transitions from period to period using the balance equations $I_t = Q_t + I_{t-1} - D_t$, for $t = 1, \ldots, T$ (we assume $I_0 = 0$). That is, the inventory at the end of a period equals the inventory at the end of the prior period, plus the amount produced in the period, minus the period's demand. In order to track fixed order costs, we define the binary variable y_t, which equals one if an order is placed in period t, and zero otherwise, for $t = 1, \ldots, T$. We assume that production or procurement capacity is unlimited in any period, and that production in period t is delivered at the beginning of period t, for $t = 1, \ldots, T$ (or, equivalently, that production decisions are offset from batch deliveries by some fixed lead time).

The planner wishes to minimize total fixed order, variable procurement, and inventory holding costs incurred over the planning horizon while meeting all demands on time (i.e., without any shortages, which is equivalent to assuming that the end-of-period inventory must be nonnegative in each period). We can thus formulate the ELSP as follows:

$$\text{[ELSP]} \quad \text{Minimize} \quad \sum_{t=1}^{T} \{S_t y_t + C_t Q_t + H_t I_t\} \tag{1.11}$$

$$\text{Subject to} \quad I_t = Q_t + I_{t-1} - D_t, \quad t = 1, \ldots, T, \tag{1.12}$$

$$Q_t \le M_t y_t, \quad t = 1, \ldots, T, \tag{1.13}$$

$$Q_t, I_t \ge 0, \quad t = 1, \ldots, T, \tag{1.14}$$

$$y_t \in \{0, 1\}, \quad t = 1, \ldots, T. \tag{1.15}$$

The objective function (1.11) minimizes the sum of fixed order, variable procurement, and inventory holding costs, while the first constraint set (1.12) corresponds to the inventory balance requirements discussed previously. The second constraint set (1.13) forces production in a period to zero if no order is placed, and permits production to take a positive value up to M_t in period t if an order is placed. The parameter M_t corresponds to a large positive number that effectively ensures that no capacity limit exists (we can set $M_t = \sum_{\tau=t}^{T} D_\tau$ without loss of optimality). The third constraint set (1.14) ensures nonnegativity of order quantities and inventory levels, while the final constraint set (1.15) requires each fixed order variable to take a value of zero or one.

Despite our formulation of the ELSP as a mixed-integer linear program, the problem possesses special structure that permits solving it very efficiently, even for large values of T. The most important property possessed by the model is the so-called zero-inventory-ordering (ZIO) property, which says that an optimal solution exists such that $Q_t \times I_{t-1} = 0$ for all values of t. This implies that an optimal solution exists such that if an order is placed in period t ($Q_t > 0$), then no inventory is held over from period $t-1$ ($I_{t-1} = 0$); similarly, an optimal solution exists such that if we hold inventory at the end of period t ($I_t > 0$), then no production occurs in period $t+1$ ($Q_{t+1} = 0$), and this holds for all values of t. This ensures that we will find an optimal solution if we confine ourselves to solutions of the form $Q_t = \sum_{\tau=t}^{s} D_\tau$ for each t, with s equal to some time period index greater than or equal to $t-1$ (using the convention $\sum_{\tau=t}^{t-1} D_\tau = 0$). The values of Q_t must of course be compatible in forming a solution for the T-period problem, i.e., if $Q_t = \sum_{\tau=t}^{s} D_\tau$ with $s > t$, then we require $Q_\tau = 0$ for $\tau = t+1, \ldots, s$ and $Q_{s+1} > 0$ (assuming $s < T$ and positive demand in every period). A solution method that implicitly considers all solutions of this form can be obtained by using a shortest path graph containing $T+1$ nodes in which an arc is created from node t to $s+1$ (for each t and all $s \geq t$) that accounts for all costs incurred when using the order in period t to satisfy all demands in periods t through s inclusive. The resulting shortest path graph contains $\mathscr{O}(T^2)$ arcs, which implies a worst-case complexity[1] of $\mathscr{O}(T^2)$ for solving this problem. The definition of the ELSP and an $\mathscr{O}(T^2)$ solution approach were first provided in [11]. Three papers subsequently appeared that showed how to solve the problem more quickly, in $\mathscr{O}(T \log T)$ time when costs vary with time, and in $\mathscr{O}(T)$ time under non-increasing marginal costs[2] (see [1, 4], and [10]). The ELSP formulation (1.11)–(1.15) and basic solution approaches we have described will form the basis for the problems we will discuss in Chaps. 5 and 6.

1.2.4 The Knapsack Problem (KP)

The knapsack problem is perhaps the simplest combinatorial *demand selection* problem, and is also one of the easiest operations research problems to explain to the lay-person. This problem considers a single resource (the knapsack) with limited (positive) capacity. The problem considers a set of items, a subset of which may be inserted in the knapsack. Each item has a value, and the decision maker wishes to maximize the value of the items inserted in the knapsack. Of course, if all items in the set fit in the knapsack, the problem is trivial and the value is maximized by including all of the items with positive value (we can immediately eliminate all items with non-positive value from consideration without loss of generality). We therefore consider problems in which the capacity consumption of the set of items under consideration exceeds the knapsack's capacity. We will refer to items as demands,

---

[1] The notation $\mathscr{O}(T^2)$ implies that some constant K exists such that as T increases, the number of steps required to solve the problem is bounded by KT^2.

[2] More specifically, non-increasing marginal costs imply $C_t + H_t \geq C_{t+1}$ for $t = 1, \ldots, T-1$, i.e., given that orders are placed in periods s and t with $s > t$, then satisfying a unit of demand in period s or later is at least as cheap when using production in period s as it is when using production in period t.

as each item contains an inherent demand for the knapsack's capacity; similarly, we will refer to the knapsack using the generic term *resource*.

To formulate this problem, let J denote a set of n demands, indexed by j, and define x_j as a binary variable equal to one if demand j is allocated to the resource (i.e., selected), and zero otherwise. Let D_j denote the capacity consumption associated with demand j, and let b denote the total capacity of the resource (we assume that the resource capacity and demand consumption are measured in consistent units using a single dimension). Letting R_j denote the value associated with demand j, we formulate the KP as follows:

$$[\text{KP}] \quad \text{Maximize} \quad \sum_{j=1}^{n} R_j x_j \tag{1.16}$$

$$\text{Subject to} \quad \sum_{j=1}^{n} D_j x_j \leq b, \tag{1.17}$$

$$x_j \in \{0, 1\}, \quad j = 1, \ldots, n. \tag{1.18}$$

The objective of KP (1.16) maximizes the total value of selected demands, the single constraint (1.17) enforces the resource capacity limit, and the final constraint set (1.18) ensures that every demand is either selected or rejected. Although the recognition version[3] of the KP is $\mathcal{NP}$-complete (see [6]), the KP can be solved in pseudopolynomial time in the number of demands and the resource capacity (a dynamic programming approach can be applied to solve the problem in $\mathcal{O}(nb)$ time). The solution of the continuous relaxation of KP (where each x_j is permitted to take any value on the interval $[0, 1]$) is quite intuitive, as it nicely captures the trade-off between a demand's value and resource capacity consumption (note that the continuous version is equivalent to being able to select a portion of any demand). This solution works by sorting demands in nonincreasing order of their value-to-capacity-consumption ratios, i.e., R_j/D_j, and allocating them to the resource as long as capacity permits. After sorting demands in this order, let k denote the index of the unique item[4] such that $\sum_{j=1}^{k-1} D_j \leq b$ and $\sum_{j=1}^{k} D_j > b$. Then an optimal solution sets $x_j = 1$ for $j = 1, \ldots, k-1$, $x_k = (b - \sum_{j=1}^{k-1} D_j)/D_k$, and $x_j = 0$ for $j = k+1, \ldots, n$ (unless $\sum_{j=1}^{k-1} D_j = b$, in which case we simply have $x_k = 0$).

The structure of this optimal solution is quite useful with respect to the original binary version of the problem, as the adjusted solution in which $x_k = 0$ (and the values of all other x_j variables are unchanged) turns out to be feasible for KP. Under mild assumptions on the distributions of the values and resource consumption

[3] In the language of complexity theory, the recognition version of an optimization problem with a maximization objective asks the question "Does a feasible solution exist with objective function value at least equal to K for some constant K?" Thus, the recognition version of the problem always has a yes/no answer (see [6]).

[4] We assume uniqueness of R_j/D_j ratios, as items with identical values may be combined into one item in the continuous version of the problem.

parameters, as well as on the resource capacity as the number of items increases, it is possible to show that the resulting feasible solution is asymptotically optimal for the KP as the number of items and the knapsack capacity increase (see [5]). Later, in Chap. 8, we will encounter a generalized version of the KP that considers flexibility in demand sizes, where each value of D_j becomes a decision variable whose value must fall between some lower and upper bounds (and where the quantity of demand satisfied depends on the chosen value of D_j).

1.2.5 The Generalized Assignment Problem (GAP)

The generalized assignment problem (GAP) arises in many production and logistics settings in which *jobs* must be assigned to *resources*. Examples include the assignment of jobs to a production machine, or of customer shipments to a delivery truck. In many contexts these jobs correspond to specific customer requirements or demands that must be fulfilled. Thus, we use the term demands in place of jobs in describing the GAP. As with the newsvendor problem, the GAP has no time dimension, which means that we can view this as a one-period problem. We consider a set I of m resources indexed by i that can be used to fulfill demands, along with a set J of n demands indexed by j. Resource $i \in I$ has a capacity of b_i units, while demand $j \in J$ requires D_{ij} units of capacity if it is fulfilled using resource i. Processing demand j on resource i costs c_{ij} dollars. The GAP requires assigning each demand $j \in J$ to some resource $i \in I$. A demand may not be assigned to multiple resources, i.e., no splitting of a demand's processing requirements among multiple resources is permitted. The planner wishes to assign all demands at minimum total cost while obeying each resource's capacity limit. Defining x_{ij} as a binary variable equal to one if demand j is assigned to resource i, and zero otherwise, we can formulate the GAP as follows:

$$\text{[GAP]} \quad \text{Minimize} \quad \sum_{i=1}^{m}\sum_{j=1}^{n} c_{ij}x_{ij} \tag{1.19}$$

$$\text{Subject to} \quad \sum_{j=1}^{n} D_{ij}x_{ij} \leq b_i, \quad i = 1, \ldots, m, \tag{1.20}$$

$$\sum_{i=1}^{m} x_{ij} = 1, \qquad j = 1, \ldots, n, \tag{1.21}$$

$$x_{ij} \in \{0, 1\}, \qquad i = 1, \ldots, m, \; j = 1, \ldots, n. \tag{1.22}$$

The objective of the GAP (1.19) minimizes total assignment costs, while the first constraint set (1.20) ensures that the resource requirements for demands assigned to resource i do not exceed the resource capacity b_i, for all $i \in I$. The second constraint set (1.21) requires assigning each demand $j \in J$ to some resource $i \in I$, while the final constraint set (1.22) ensures that each assignment variable takes a value of zero or one. Later in Chaps. 7 and 8 we will explore a more general version of the GAP that permits flexibility in satisfying demand requirements. For an early definition of the GAP, please see [3].

1.2.6 The Facility Location Problem (FLP)

The facility location problem (FLP) we next present generalizes two of the models we have already discussed: the ELSP and the GAP. Although the model originated in contexts requiring location decisions for physical facilities, we can view it as a more generic fixed-charge assignment or bipartite network flow problem. In particular, we can view the FLP as a generalization of the GAP in which a fixed-charge is incurred when using any resource. Thus, if S_i denotes the fixed charge for using resource i, and y_i denotes a binary variable equal to one if resource i is utilized (and zero otherwise), then we can formulate this more general version of the GAP as follows:

$$\text{[FLP]} \quad \text{Minimize} \quad \sum_{i=1}^{m} S_i y_i + \sum_{i=1}^{m}\sum_{j=1}^{n} c_{ij} x_{ij} \tag{1.23}$$

$$\text{Subject to} \quad \sum_{j=1}^{n} D_j x_{ij} \le b_i y_i, \quad i = 1, \ldots, m, \tag{1.24}$$

$$\sum_{i=1}^{m} x_{ij} = 1, \quad j = 1, \ldots, n, \tag{1.25}$$

$$x \in \Omega, \tag{1.26}$$

$$y_i \in \{0, 1\}, \quad i = 1, \ldots, m. \tag{1.27}$$

The objective of the FLP (1.23) minimizes the sum of resource fixed costs plus demand assignment costs. The first constraint set (1.24) differs from (1.20) because the demand is independent of the facility (hence the D_j instead of D_{ij}) and in the extra y_i term multiplying the capacity on the right-hand side. This constraint thus only permits assigning demands to a resource if the associated fixed cost is absorbed (and $y_i = 1$). The third constraint set (1.26) differs from our formulation of the GAP as we now require that each variable x_{ij} is a member of some $m \times n$ dimensional set Ω. When $\Omega = [0, 1]^{m\times n}$, then we permit splitting each demand among multiple resources by allowing the assignment (x_{ij}) variables to be continuous. When $\Omega = \{0, 1\}^{m\times n}$, then this version of the problem is known as the FLP with single-sourcing requirements; strictly speaking, the FLP with single sourcing generalizes the GAP, and the formulation of the FLP when $\Omega = [0, 1]^{m\times n}$ is a relaxation of the single-sourcing version of the problem (note that when capacities are unlimited, this distinction is not necessary, as an optimal single-sourcing solution is guaranteed to exist). The remaining constraint set (1.27) imposes binary restrictions on the new fixed-charge (y_i) variables.

Observe that in addition to generalizing the ELSP and GAP, when the assignment constraints (1.25) are relaxed (either by omitting them or by changing the equality to a less than or equal to relation) and $m = 1$ (only a single resource exists with $S_1 = 0$), then the FLP reduces to the KP if the c_{1j} values are permitted to be negative (and $-c_{1j}$ is equal to R_j).

To see how the ELSP serves as a special case of FLP, observe that we can reformulate the ELSP as follows. We first define x_{ts} as the percentage of demand in period s that is satisfied using procurement in period t, for all $t = 1, \ldots, T$ and $s = t, \ldots, T$. In terms of the ELSP formulation, we have $Q_t = \sum_{s=t}^{T} D_s x_{ts}$ and $I_t = \sum_{\tau=1}^{t} \sum_{s=\tau}^{T} D_s x_{\tau s} - \sum_{\tau=1}^{t} D_\tau$. We next define c_{ts} as the cost (procurement plus holding) to satisfy demand in period s using production in period t, i.e., $c_{ts} = (C_t + \sum_{\tau=t}^{s-1} H_\tau) D_s$. We can then formulate the ELSP equivalently as follows:

$$\text{[FELSP]} \quad \text{Minimize} \quad \sum_{t=1}^{T} S_t y_t + \sum_{t=1}^{T} \sum_{s=t}^{T} c_{ts} x_{ts} \tag{1.28}$$

$$\text{Subject to} \quad \sum_{s=t}^{T} D_s x_{ts} \leq M_t y_t, \quad t = 1, \ldots, T, \tag{1.29}$$

$$\sum_{t=1}^{s} x_{ts} = 1, \quad s = 1, \ldots, T, \tag{1.30}$$

$$x_{ts} \geq 0, \quad t = 1, \ldots, T, \; s = t, \ldots, T, \tag{1.31}$$

$$y_t \in \{0, 1\}, \quad t = 1, \ldots, T. \tag{1.32}$$

The FELSP formulation is equivalent to the ELSP formulation, and is a special case of the FLP in which a demand may be assigned to only a subset of the resources, and resource capacities are effectively unlimited (in this case, the resources correspond to period orders or production setups). In addition, the FELSP formulation implicitly assumes that the size of each demand s is independent of the resource to which it is assigned (hence the single s index for D_s). By replacing each M_t with a finite value b_t, we obtain the capacitated version of the ELSP. Note that the uncapacitated version of the FLP is obtained by replacing each b_i in (1.24) with a big-M_i value (e.g., $M_i = \sum_{j=1}^{n} D_{ij}$ for each $i = 1, \ldots, m$). The uncapacitated version of the FLP (and, therefore, the FELSP formulation) has the property that an optimal solution exists in which the x_{ij} variables are binary, i.e., a single-sourcing solution is optimal even though it is not explicitly required.

References

1. Aggarwal A, Park J (1993) Improved Algorithms for Economic Lot Size Problems. Operations Research 41(3):549–571

2. Anton H (1988) Calculus. Wiley, New York
3. De Maio A, Roveda C (1971) An All Zero-One Algorithm for a Certain Class of Transportation Problems. Operations Research 19(6):1406–1418
4. Federgruen A, Tzur M (1991) A Simple Forward Algorithm to Solve General Dynamic Lot Sizing Models with n Periods in $\mathcal{O}(n \log n)$ or $\mathcal{O}(n)$ Time. Management Science 37(8):909–925
5. Frieze A, Clarke M (1984) Approximation Algorithms for the m-Dimensional 0–1 Knapsack Problem: Worst-Case and Probabilistic Analyses. European Journal of Operational Research 15(1):100–109
6. Garey M, Johnson D (1979) Computers and Intractability. W.H. Freeman and Company, New York
7. Hadley G, Whitin T (1961) An Optimal Final Inventory Model. Management Science 7(2):179–183
8. Harris F (1913) How Many Parts to Make at Once. Factory Magazine Management 10:135–136, 152
9. Silver E, Pyke D, Peterson R (1998) Inventory Management and Production Planning and Scheduling, 3rd edn. Wiley, New York
10. Wagelmans A, van Hoesel S, Kolen A (1992) Economic Lot Sizing: An $\mathcal{O}(n \log n)$ Algorithm That Runs in Linear Time in the Wagner-Whitin Case. Operations Research 40(S1):S145–S156
11. Wagner H, Whitin T (1958) Dynamic Version of the Economic Lot Size Model. Management Science 5(1):89–96

Chapter 2
Production and Inventory Planning Models with Demand Shaping

Abstract This chapter considers the state of prior literature on operations models that account for demand flexibility. In particular, we focus on generalizations of the models discussed in Chap. 1 that treat demands as decision variables. These generalizations typically involve pricing models, and they provide a foundation for the models we will study in subsequent chapters.

2.1 EOQ Models with Pricing

Whitin [37] provided a seminal paper on inventory control and pricing under the EOQ model assumptions (this paper also considers a single-period stochastic problem, which we discuss in the following section). This model generalizes cost equation (1.4) to account for price-dependent demand and subsequently maximizes profit per unit time instead of cost. In order to do this, a linear price–demand function is assumed that takes the form

$$D = \beta - \alpha p, \tag{2.1}$$

where α and β are scalars (which are typically assumed to be positive) and p denotes price. Using this demand function (2.1) along with cost Eq. (1.4), we can write the average profit per unit time as a function of p, denoted as $\Pi(p)$, as

$$\Pi(p) = (p - C)(\beta - \alpha p) - \sqrt{2S(\beta - \alpha p)H}. \tag{2.2}$$

Letting $Q^*(p) = \sqrt{2S(\beta - \alpha p)/H}$, Arcelus and Srinivasan [3] provide the following form of the stationary-point solution for the optimal price[1]:

$$p^* = \frac{1}{2}\left[\frac{\beta}{\alpha} + C + \frac{S}{Q^*(p^*)}\right], \tag{2.3}$$

[1] One can verify that the second derivative of $\Pi(p)$ with respect to p is strictly increasing in p; thus the profit function $\Pi(p)$ is either convex in p for all $p \geq 0$, in which case an optimal extreme point solution exists (i.e., either $p^* = 0$ or $p^* = \beta/\alpha$), or $\Pi(p)$ is concave on some interval $[0, \tilde{p}]$ and convex for $p \geq \tilde{p}$. In the latter case, either an extreme solution or the stationary-point solution (2.3) is optimal (assuming such a stationary point exists).

J. Geunes, *Demand Flexibility in Supply Chain Planning*, SpringerBriefs in Optimization, DOI 10.1007/978-1-4419-9347-2_2,

where

$$Q^*(p^*) = \sqrt{\frac{2S(\beta - \alpha p^*)}{H}}. \tag{2.4}$$

This generalized version of the EOQ model permits selecting the *optimal* demand level when demand is price-dependent (for simplicity, we have considered a linear price–demand function, although more complex functions have been explored). Numerous additional generalizations of this basic model have been addressed in the literature, including problems with more general demand functions ([23, 25, 31]), quantity discounts ([1, 8, 9]), and investment and storage constraints ([7, 24]).

2.2 The Newsvendor Problem with Pricing and Demand Shaping

As noted in the previous section, Whitin [37] first considered a single period inventory problem with pricing under demand uncertainty. This problem used a marginal analysis with a unit profit for items sold and a unit loss associated with excess inventory remaining after demand is realized. Assuming that expected demand is a linear function of the profit margin and that demand is uniformly distributed between zero and twice the expected demand, Whitin [37] derived an expression for the optimal profit margin. The demand model used by Whitin [37] is effectively a *multiplicative* model, as both the expected value and the variance of the demand distribution depend on price. This is in contrast with the *additive* model, where demand is expressed as a deterministic function of price plus a random error term, which is independent of price. That is, letting $D(p, \varepsilon)$ denote the demand function, where ε is a random variable, then in an additive model, $D(p, \varepsilon) = y(p) + \varepsilon$, where $y(p)$ is a deterministic function of price and ε is a random variable with mean μ and variance σ^2. Under a multiplicative model, we have $D(p, \varepsilon) = y(p)\varepsilon$. The form of the demand model assumed fundamentally affects both the quantitative and qualitative results of the model. We will first briefly illustrate the application of a simple additive demand model and then discuss more general work that subsumes both the additive and multiplicative cases.

We initially consider the basic newsvendor problem under normal demand discussed in Chap. 1. Consider a generalized version of the expected profit equation (1.10) in which the expected demand μ_D is price dependent. In particular, assume that $\mu_D = \alpha - \beta p$. In this case, Eq. (1.10) becomes

$$\Pi_n(Q^*, p) = (p - C)(\alpha - \beta p) - K(z^*)\sigma_D. \tag{2.5}$$

This profit equation is concave in p with stationary-point solution (and therefore optimal solution) $p_a^* = (1/2)(C + (\alpha/\beta))$, where the subscript a corresponds to the additive case. Observe that for this additive demand model, the optimal price depends only on the profit margin $(p - C)$ and the expected demand, i.e., the optimal price is the same as that for the zero-variance (risk-free) case. The same is not true in a multiplicative model.

To illustrate this, suppose that, in addition, σ_D is a linear function of price, i.e., $\sigma_D = (\alpha - \beta p)\sigma$. This is equivalent to a multiplicative model in which $y(p) = (\alpha - \beta p)$ and ε is normally distributed with expected value one and variance σ^2. In this special case, the expected profit equation remains concave in p, and the stationary-point optimal solution becomes $p_m^* = (1/2)(C + (\alpha/\beta) + K(z^*)\sigma) = p_a^* + (K(z^*)\sigma/2)$, where the subscript m corresponds to the multiplicative case. Thus, the price in the multiplicative case equals the risk-free (additive) price plus a premium for the way in which price affects uncertainty.

Petruzzi and Dada [29] provide an excellent, general, and detailed analysis of the newsvendor problem with pricing. We next summarize their main results, which unify the treatment of the additive and multiplicative cases. These results build on the foundations provided in [13, 21, 28, 37, 38], and [39]. Petruzzi and Dada [29] consider an expected profit function of the form

$$\Pi(Q, p) = (p - C)\mathrm{E}\big[\mathrm{Sales}(\zeta, p)\big] - (C + H)\mathrm{E}\big[\mathrm{Leftovers}(\zeta, p)\big] - B\mathrm{E}\big[\mathrm{Shortages}(\zeta, p)\big], \tag{2.6}$$

where a one-to-one correspondence exists between the variable ζ and the order quantity Q at any price. The relationship between Q and ζ depends on the form of the demand function. In the additive case, ζ is defined using $\zeta = Q - y(p)$. In the multiplicative case, ζ is defined using $\zeta = Q/y(p)$.

Petruzzi and Dada [29] show that, for both the additive and multiplicative demand cases, ζ can be written as

$$\zeta = \mu + \mathrm{SF}\sigma, \tag{2.7}$$

where SF is defined in [32] as the safety factor, which is the number of standard deviations by which the order quantity differs from the expected value of demand, i.e.,

$$\mathrm{SF} = \frac{Q - \mathrm{E}[D(p, \varepsilon)]}{\mathrm{SD}[D(p, \varepsilon)]}, \tag{2.8}$$

where $\mathrm{SD}[D(p, \varepsilon)]$ is the standard deviation of $D(p, \varepsilon)$. They then define the *base price*, $p_B(\zeta)$ as the price that maximizes the expected contribution to profit from sales, i.e., the price that maximizes $(p - C)\mathrm{E}[\mathrm{Sales}(\zeta, p)]$ (note that $p_B(\zeta)$ maximizes the risk-free profit, i.e., expected profit when variance equals zero). Their first main result shows that for both the additive and multiplicative cases, for a given ζ, $p_B(\zeta)$ is determined by the unique value of p satisfying

$$p = C - \frac{\mathrm{E}[\mathrm{Sales}(\zeta, p)]}{\partial \mathrm{E}[\mathrm{Sales}(\zeta, p)]/\partial p}. \tag{2.9}$$

The second main result states that the optimal price in both the multiplicative and additive cases is bounded from below by $p_B(\zeta)$ for any given ζ. This implies that for both cases we can view the optimal price as the optimal base price plus a premium. In the additive case this premium equals zero because, for any given ζ, the expected leftover and shortage costs are independent of price. In the multiplicative case, the premium depends on the impact the price has on expected holding and shortage costs. Petruzzi and Dada [32] provide functional forms for the optimal price in both the multiplicative and additive cases, as well as methods to determine the optimal corresponding order quantity.

In addition to shaping demand through pricing, a few papers have considered different dimensions of demand flexibility in a stochastic demand setting. Petruzzi and Monahan [30] consider a fashion goods context with a primary and secondary market, where the decision maker must determine the optimal time at which to move a good from the primary to the secondary market. Carr and Duenyas [5] consider a production system with two classes of demand, each of which has a Poisson distributed arrival rate. Type 1 demands are called make-to-stock demands, and Type 2 demands are called make-to-order demands. Type 1 demands result in a shortage cost if a demand occurs and stock is depleted, while Type 2 demands may be accepted or rejected, although those accepted demands are made-to-order. This work provides an optimal production and order acceptance policy for this problem class. Carr and Lovejoy [6] consider a single-period newsvendor-type problem in which a number of prioritized demand portfolios are available, and a single resource with random capacity may be used to satisfy demands. The decision maker must determine the amount of demand to select within each portfolio in order to maximize expected profit.

2.3 Lot Sizing with Pricing

The earliest work on integrating pricing in the ELSP appears to be that of Thomas [33]. His work generalized the ELSP to incorporate the dependence of demand in each period on price. In this model, price may vary from period to period, and demand in period t depends on the price in period t, p_t, according to the function $D_t(p_t)$. This generalization of the ELSP can be formulated as follows:

$$[\text{ELSP}'] \quad \text{Maximize} \quad \sum_{t=1}^{T}\left\{p_t D_t(p_t) - S_t y_t - C_t Q_t - H_t I_t\right\} \tag{2.10}$$

$$\text{Subject to} \quad I_t = Q_t + I_{t-1} - D_t(p_t), \quad t = 1, \ldots, T, \tag{2.11}$$

$$Q_t \leq M_t y_t, \quad t = 1, \ldots, T, \tag{2.12}$$

$$Q_t, I_t, p_t \geq 0, \quad t = 1, \ldots, T, \tag{2.13}$$

$$y_t \in \{0, 1\}, \quad t = 1, \ldots, T. \tag{2.14}$$

The solution approach relies on the fact that for any given price vector, the problem reduces to the ELSP, and the zero-inventory-ordering (ZIO) property continues to hold. This implies that the shortest path solution approach discussed in Chap. 1 may still be applied in principle, although it becomes an acyclic longest path problem in which arcs are assigned profits instead of costs. If, for example, production in period t satisfies demand in periods t through s, then the profit on the arc from node t to node $s+1$ is obtained by solving the following pricing subproblem PSP where, with a slight abuse of notation, $H_{t,\tau} = \sum_{u=t}^{\tau-1} H_u$:

$$\text{[PSP]} \quad \text{Maximize} \quad \sum_{\tau=t}^{s} \{(p_\tau - C_t - H_{t,\tau}) D_\tau(p_\tau)\}. \tag{2.15}$$

The PSP above decomposes by period, and its difficulty depends on the specification of the demand functions $D_t(p_t)$. If the optimal solution value to the PSP is less than or equal to the fixed order cost in period t, S_t, then the maximum arc $(t, s+1)$ profit equals zero; otherwise the arc profit equals the optimal objective function value less S_t. Thomas [33] illustrates the case in which $D_t(p_t)$ is linear in p_t for $t = 1, \ldots, T$, which implies that the PSP is easily solved using first order conditions.

Kunreuther and Schrage [22] subsequently considered the problem when price must be time-invariant, which is equivalent to adding the constraints $p_t = p_{t+1}$ for $t = 1, \ldots, T-1$, to the ELSP′ formulation. This restriction of the problem leads to a very different algorithm for solving the problem, because the problem can no longer be solved by decomposition into smaller time horizons, and the shortest path solution we described is no longer possible. However, as we know that for any given price $p = p_1 = p_2 = \cdots = p_T$, the problem again reduces to the ELSP. Thus, we know that an optimal solution exists that satisfies the ZIO property and that may be completely characterized by the sequence of order periods. That is, if we know there are ρ order periods $t_1, t_2, \ldots, t_\rho$, then the corresponding ZIO solution produces $\sum_{\tau=t_j}^{t_{j+1}-1} D_\tau(p)$ in period t_j for $j = 1, \ldots, \rho$. We will refer to a specific set of order periods $t_1, t_2, \ldots, t_\rho$ as an *order plan*. Kunreuther and Schrage [22] assume that demand in any period t is a linear function of a *price effect* function $d(p)$, which is time invariant. That is, $D_t(p) = \alpha_t + \beta_t d(p)$ for $t = 1, \ldots, T$, where α_t and β_t are nonnegative constants for each period t. For the special case in which $d(p) = -p$, we have the familiar linear price–demand function $D_t(p) = \alpha_t - \beta_t p$.

Observe that, given any price p, a vector of demands $[D(p)] = [D_1(p), \ldots, D_T(p)]$ results. The revenue associated with this demand vector equals $\sum_{t=1}^{T} p D_t(p)$. The cost associated with this vector of demands depends on the order plan utilized (we confine ourselves to the order plans defined in the previous paragraph, since we know that an optimal solution exists from among these solutions). As shown in [22], the cost of any order plan can be expressed as a linear function of the price effect $d(p)$. Assuming $d(p) = -p$, or a linear price–demand relationship in each period, this implies that the minimum cost as a function of p is a piecewise linear and concave function of p, where each linear segment corresponds to a specific order plan. If we can specify this piecewise linear function, or *envelope*, then it is possible to evaluate the maximum profit associated with every candidate order plan contained in an optimal solution. In this case, for each segment of the piecewise linear function, we can compute the optimal price for the given order plan. Specifying this piecewise linear function is not trivial, however.

Kunreuther and Schrage [22] suggest a heuristic approach that assumes the optimal price must fall on some interval $[p_L, p_U]$. We next briefly sketch the way this heuristic works. First, recall that for any fixed price, the problem reduces to an ELSP. Thus, we can initially solve the problem at the price p_L, which requires solving an instance of the ELSP. This solution provides an optimal order plan at the price p_L. The cost of this order plan is linear in price, and the associated line must form a segment of the piecewise linear concave envelope. Given this order plan, we next determine the price that maximizes profit when restricting ourselves to this particular order plan. If this price differs from p_L, then we can solve the ELSP corresponding to this new price. If the optimal order plan differs from the previous one, then we have identified an additional segment of the piecewise linear concave envelope. We continue this procedure iteratively, until the price and order plan converge. Call the resulting price after convergence p_L^*. We then repeat this process using the starting price p_U, and converging to the price p_U^*. As shown in [22], the optimal price, p^*, satisfies $p_L^* \leq p^* \leq p_U^*$.

Gilbert [15] considered a special case of this model in which costs are time-invariant and $D_t(p) = \beta_t d(p)$. For this case, he showed that the piecewise linear concave envelope has at most $\mathcal{O}(T)$ segments, and that these segments can be identified in polynomial time. Van den Heuvel and Wagelmans [35] then provided an algorithm that permits identifying the entire piecewise linear concave envelope for the general case defined in [22]. Beginning with the solutions p_L^* and p_U^*, they show how to identify whether an unidentified segment of the piecewise linear concave envelope exists by solving the problem at the intersection of the lines corresponding to the optimal order plans at the prices p_L^* and p_U^*. If a new line segment is identified, and its optimal price is also identified, this permits eliminating part of the interval of uncertainty[2] between p_L^* and p_U^*. This can then be repeated for any remaining intervals of uncertainty. Because the number of order plans is finite, the procedure is finite and must converge to an optimal price.

Gilbert [16] considered a multiple product lot sizing problem with shared but time-invariant production capacities and a time-invariant price for each good. Deng and Yano [10] and Geunes, Merzifonluoğlu, and Romeijn [14] subsequently considered the integrated pricing and lot sizing problem with production capacities. Merzifonluoğlu, Geunes, and Romeijn [27] considered a class of aggregate planning problems in which capacities, prices, and subcontracting levels served as decision variables.

2.4 Knapsack Problems with Nonlinear Objectives

This section describes a class of continuous knapsack problems in which a set J of n demands exists, as in our discussion of knapsack problems in Chap. 1. In this class

[2]We define an interval of uncertainty as an interval which is known to contain the optimal price, although the precise value of the optimal price remains unknown.

of knapsack problems, however, the variable x_j no longer corresponds to a binary variable that determines whether or not demand j is selected. Instead, x_j denotes a variable corresponding to the percentage of some maximum level, D_j, at which demand j may be satisfied. For example, D_j might correspond to the maximum level of sales effort that may be applied in a market j or the maximum amount of advertising expenditures that may be dedicated to product j. In each of these cases, the activity level consumes part of a finite resource with capacity b (in the former case this resource may be a salesperson's time in a period, while in the latter this resource may be a limited budget).

Associated with demand j is a revenue function $R_j(x_j)$, which depends on the activity level for demand j. We can formulate this nonlinear revenue maximizing knapsack problem as follows:

$$\text{[NLKP]} \quad \text{Maximize} \quad \sum_{j=1}^{n} R_j(x_j) \tag{2.16}$$

$$\text{Subject to} \quad \sum_{j=1}^{n} D_j x_j \leq b, \tag{2.17}$$

$$0 \leq x_j \leq 1, \quad j = 1, \ldots, n. \tag{2.18}$$

Clearly if each $R_j(\cdot)$ function is linear in x_j, then NLKP corresponds to the continuous relaxation of the knapsack problem KP defined in Chap. 1. If each $R_j(\cdot)$ function is convex in x_j, then an extreme point optimal solution also exists (as in the relaxation of KP). It is straightforward to show that extreme point solutions for NLKP contain at most one x_j variable that takes a value strictly between zero and one (moreover, for such extreme solutions, the resource capacity constraint (2.17) is tight). Using this fact, [4] provides a pseudopolynomial time algorithm under convex revenue functions that runs in $\mathcal{O}(Un^2b)$ time in the worst case, where U denotes the maximum value of D_j over all $j \in J$ (assuming that b and all D_j are integer). When all revenue functions are concave, the NLKP is a convex program and can therefore be solved using standard nonlinear optimization solvers. More general so-called S-curve return functions are considered in [2] and [17]. These functions arise in numerous marketing contexts such as advertising, where small levels of investment provide increasing returns to scale, and larger investment levels lead to decreasing returns to scale. Such S-curve functions are convex from zero to an inflection point, and then are concave thereafter. Analysis of the special structure of these revenue functions leads to pseudopolynomial time solution methods (see [2, 17]).

Additional classes of generalized knapsack problems with demand flexibility will arise in our study of decomposition methods for assignment and location models in Chap. 8. Moreover, in our analysis of EOQ models in Chap. 3 and in our discussion of newsvendor models in Chap. 4, several interesting nonlinear and nonseparable knapsack problems will arise.

2.5 Location and Assignment Problems with Flexible Demand

Location theory has been well studied in the economics and operations research literature under a number of assumptions. Much of the literature on location theory with price effects applies game-theoretic analysis in competitive settings. This body of literature simultaneously considers the objectives of multiple competing organizations, each of which wishes to maximize its profit based on its location and market-supply decisions. A discussion of the models and approaches for this class of problems may be found in [11] and [34]. The models we consider, and which are most relevant to the work considered throughout this book, are more appropriate for a single firm who is a monopolist, and thus wishes to make location decisions based on response to a price–demand curve for its product (and independent of other firms' decisions).

Wagner and Falkson [36] provided perhaps the earliest model for a facility location problem facing a single monopolistic producer of a good with price-sensitive demand. This model considered the location of public facilities under the maximization of social welfare and several different assumptions on the level of service that must be provided to customers. Hansen and Thisse [19] then provided a model for a private firm seeking to simultaneously determine price and location decisions in order to maximize profit when demand is price-dependent. Erlenkotter [12] generalized their approach to account for the profit maximization objectives of private and public firms within a single model. He provided a heuristic algorithmic approach based on Lagrangian relaxation and explicitly considered situations in which the revenue in a customer market is a quadratic function of price. The models we have discussed thus far permit charging different prices to individual markets, where the optimal price in a market depends on which facility serves the market at optimality. Hansen, Thisse, and Hanjoul [20] modeled the problem when the delivered price must be the same for all markets. Hanjoul et al. [18] later provided models that allowed different methods of consistent pricing among customers (that is, they considered the case in which the delivered price is the same for all customer markets, as well as the case in which all customers pay the same *mill price*, i.e., the price before bearing transportation costs from the supply point).

Although work on the generalized assignment problem (GAP) with pricing is quite limited, a rich set of models exists in the marketing literature for determining the optimal amount of limited salesforce effort to exert in different territories (see, e.g., [26, 40]). In these models, a salesperson's time corresponds to a limited resource, and sales territories must be assigned to sales personnel. Given an assignment of territories to a salesperson, the time the salesperson spends in each territory must also be determined, where the sales response (or revenue function, as in the NLKP) in a territory is a nonlinear function of the time spent in the territory (or the effort exerted). Thus, the sales level within each territory (i.e., the demand) effectively serves as a decision variable that is determined via the level of sales effort.

References

1. Abad P (1988) Determining Optimal Selling Price and Lot Size When the Supplier Offers All-Unit Quantity Discounts. Decision Sciences 19(3):622–634
2. Ağralı S, Geunes J (2009) Solving Knapsack Problems with S-Curve Return Functions. European Journal of Operational Research 193:605–615
3. Arcelus F, Srinivasan G (1998) Ordering Policies under One Time Only Discount and Price Sensitive Demand. IIE Transactions 30:1057–1064
4. Burke G, Geunes J, Romeijn H, Vakharia A (2008) Allocating Procurement to Capacitated Cuppliers with Concave Quantity Discounts. Operations Research Letters 36:103–109
5. Carr S, Duenyas I (2000) Optimal Admission Control and Sequencing in a Make-to-Stock/Make-to- Order Production System. Operations Research 48(5):709–720
6. Carr S, Lovejoy W (2000) The Inverse Newsvendor Problem: Choosing an Optimal Demand Portfolio for Capacitated Resources. Management Science 46(7):912–927
7. Cheng T (1990) An EOQ Model with Pricing Consideration. Computers and IE 18(4):529–534
8. Dada M, Srikanth K (1987) Pricing Policies for Quantity Discounts. Management Science 33(10):1247–1252
9. Das C (1984) A Unified Approach to the Price-Break Economic Order Quantity (EOQ) Problem. Decision Sciences 15(3):350–358
10. Deng S, Yano C (2006) Joint Production and Pricing Decisions with Setup Costs and Capacity Constraints. Management Science 52(5):741–756
11. Eiselt H, Laporte G, Thisse J (1993) Competitive Location Models: A Framework and Bibliography. Transportation Science 27(1):44–54
12. Erlenkotter D (1977) Facility Location with Price-Sensitive Demands: Private, Public, and Quasi-Public. Management Science 24(4):378–386
13. Ernst R (1970) A Linear Inventory Model of a Monopolistic Firm. PhD Dissertation, Department of Economics, University of California, Berkeley, CA
14. Geunes J, Merzifonluoğlu Y, Romeijn H (2009) Capacitated Procurement Planning with Price-Sensitive Demand and General Concave Revenue Functions. European Journal of Operational Research 194:390–405
15. Gilbert S (1999) Coordination of Pricing and Multiple-Period Production for Constant Priced Goods. European Journal of Operational Research 114:330–337
16. Gilbert S (2000) Coordination of Pricing and Multiple-Period Production Across Multiple Constant Priced Goods. Management Science 46:1602–1616
17. Ginsberg W (1974) The Multiplant Firm with Increasing Returns to Scale. Journal of Economic Theory 9:283–292
18. Hanjoul P, Hansen P, Peeters D, Thisse J (1990) Uncapacitated Plant Location under Alternative Spatial Price Policies. Management Science 36(1):41–57
19. Hansen P, Thisse J (1977) Multiplant Location for Profit Maximization. Environment and Planning A 9:63–73
20. Hansen P, Thisse J, Hanjoul P (1980) Simple Plant Location under Uniform Delivered Pricing. European Journal of Operational Research 6:94–103
21. Karlin S, Carr C (1962) Prices and Optimal Inventory Policy. In Studies in Applied Probability and Management Science (Arrow K, Karlin S, Scarf H (eds.)) Stanford University Press, Stanford, CA, 159–172
22. Kunreuther H, Schrage L (1973) Joint Pricing and Inventory Decisions for Constant Priced Items. Management Science 19(7):732–738
23. Ladany S, Sternlieb A (1974) The Interaction of Economic Ordering Quantities and Marketing Policies. AIIE Transactions 6(1):35–40
24. Lee W (1994) Optimal Order Quantities and Prices with Storage Space and Inventory Investment Limitations. Computers and IE 26(3):481–488
25. Lee W, Kim D (1993) Optimal and Heuristic Decision Strategies for Integrated Production and Marketing Planning. Decision Sciences 24(6):1203–1213

26. Lodish L (1971) CALLPLAN: An Interactive Salesman's Call Planning System. Management Science 18:4 Part II:25–40
27. Merzifonluoğlu Y, Geunes J, Romeijn H (2007) Integrated Capacity, Demand, and Production Planning with Subcontracting and Overtime Options. Naval Research Logistics 54(4):433–447
28. Mills E (1959) Uncertainty and Price Theory. The Quarterly Journal of Economics 73:116–130
29. Petruzzi N, Dada M (1999) Pricing and the Newsvendor Problem: A Review with Extensions. Operations Research 47(2):183–194
30. Petruzzi N, Monahan G (2003) Managing Fashion Goods Inventories: Dynamic Recourse for Retailers with Outlet Stores. IIE Transactions 35(11):1033–1047
31. Rosenberg D (1991) Optimal Price-Inventory Decisions-Profit vs. ROII. IIE Transactions 23(1):17–22
32. Silver E, Pyke D, Peterson R (1998) Inventory Management and Production Planning and Scheduling, 3rd edn. Wiley, New York
33. Thomas J (1970) Price-Production Decisions with Deterministic Demand. Management Science 16(11):747–750
34. Tobin R, Miller T, Friesz T (1995) Incorporating Competitors' Reactions in Facility Location Decisions: A Market Equilibrium Approach. Location Science 3(4):239–253
35. Van den Heuvel W, Wagelmans A (2006) A Polynomial Time Algorithm for a Deterministic Joint Pricing and Inventory Model. European Journal of Operational Research 170:463–480
36. Wagner J, Falskon L (1975) The Optimal Nodal Location of Public Facilities with Price-Sensitive Demand. Geographical Analysis 7(1):69–83
37. Whitin T (1955) Inventory Control and Price Theory. Management Science 2(1):61–68
38. Young L (1978) Price, Inventory and the Structure of Uncertain Demand. New Zealand Operations Research 6:157–177
39. Zabel E (1970) Monopoly and Uncertainty. The Review of Economic Studies 37:205–219
40. Zoltners A (1976) Integer Programming Models for Sales Territory Alignment to Maximize Profit. Journal of Marketing Research 13(4):426–430

Part II
Production Planning with Demand Flexibility

Chapter 3
EOQ-Type Models with Demand Selection

Abstract This chapter discusses a generalized version of the economic order quantity (EOQ). In particular, we consider a situation in which a single inventory stage must select from among a set of demand streams, those which it will satisfy. Each demand stream carries with it a constant demand rate as well as a constant revenue rate. We consider several problem variants within this class, including problems with lot size and demand rate constraints.

3.1 Unconstrained EOQ Problems with Market Choice

The problems and results discussed in this chapter are based on work previously published in [2], which contains detailed results on this problem class. We will first consider a generalization of the standard EOQ model in Sect. 3.1.1 (with an infinite production rate), and then extend the results to the case with a finite production rate in Sect. 3.1.2.

3.1.1 Standard EOQ with Market Choice

We consider a set J of n markets, indexed by j, where market j demand occurs at a constant rate of D_j units per unit time. In the analysis of the basic EOQ problem in Chap. 1, a single "market" existed and the supplier was required to serve all of this market's demand. In the EOQ problem with market choice (EOQMC), the supplier can choose whether or not to serve the demand in each market. If the supplier chooses to satisfy demand in market j, it must satisfy all of the market's demand without any shortages. We therefore define a set of binary variables, where x_j equals one if the supplier meets all demand in market j, and zero otherwise. Thus, given the values of these variables, the single stage faces a demand rate equal to

$$\sum_{j \in J} D_j x_j. \tag{3.1}$$

J. Geunes, *Demand Flexibility in Supply Chain Planning*,
SpringerBriefs in Optimization,
DOI 10.1007/978-1-4419-9347-2_3,

Under this demand rate, the optimal batch order quantity equals

$$Q^*(x) = \sqrt{\frac{2S\sum_{j \in J} D_j x_j}{H}}, \tag{3.2}$$

where x denotes an n-dimensional binary vector of x_j values. If r_j denotes the net revenue per unit of demand sold in market j (in excess of the variable cost C), then if the supplier satisfies demand in market j, the average revenue per unit time from market j equals $R_j = r_j D_j$. Thus, the average total revenue per unit time can be expressed as

$$\sum_{j \in J} R_j x_j. \tag{3.3}$$

Referring back to Eq. (1.4), we can express average profit per unit time from market selection decisions, $\Pi(x)$, as

$$\Pi(x) = \sum_{j \in J} R_j x_j - \sqrt{2SH \sum_{j \in J} D_j x_j}. \tag{3.4}$$

The supplier wishes to maximize $\Pi(x)$ over all x vectors in $\mathscr{B}^n$, the space containing all n-dimensional binary vectors. The following result, from [3], will be utilized in solving this problem as well as additional problems in this chapter and the next.

Property 3.1 *Consider the problem* $\max_{x \in \mathscr{B}^n}\{\sum_{j \in J} R_j x_j - \sqrt{\sum_{j \in J} \kappa_j x_j}\}$, *with* $\kappa_j > 0$ *for all* $j = 1, \ldots, n$. *Assuming demands are sorted in nonincreasing order of the ratio* R_j/κ_j, *if an optimal solution exists with* $x_k = 1$ *for some index* k, *then an optimal solution exists with* $x_{k-1} = 1$.

Proof Because the objective function is convex in x, this implies that an extreme point solution exists for solving the relaxation in which the requirement $x \in \mathscr{B}^n$ is replaced with $0 \le x_j \le 1$ for all $j = 1, \ldots, n$. Observe that the objective is differentiable for $x \ge 0$ everywhere except at $x = 0$. Thus, the Karush–Kuhn–Tucker (KKT) optimality conditions are *necessary* for an optimal solution (see [1]), except possibly at the point $x = 0$, where average profit per unit time equals zero. We can therefore consider all KKT points, along with the solution $x = 0$ in searching for an optimal solution. Let α_j^0 (α_j^1) denote the KKT multiplier associated with the constraint $-x_j \le 0$ ($x_j \le 1$). The associated KKT conditions can be written as

$$R_j - \frac{\kappa_j}{2\sqrt{\sum_{j \in J} \kappa_j x_j}} - \alpha_j^1 + \alpha_j^0 = 0, \quad \forall j \in J, \tag{3.5}$$

$$\alpha_0 x_j = 0, \quad \forall j \in J, \tag{3.6}$$

$$\alpha_1(1 - x_j) = 0, \quad \forall j \in J, \tag{3.7}$$

$$0 \le x_j \le 1, \quad \forall j \in J, \tag{3.8}$$

$$\alpha_j^0, \alpha_j^1 \ge 0, \quad \forall j \in J. \tag{3.9}$$

Because all extreme point solutions are binary, we only need to consider such solutions. Thus, each x_j equals zero or one. If $x_j = 1$, then we must have $\alpha_j^0 = 0$, which implies

$$\frac{R_j}{\kappa_j} \geq \frac{1}{2\sqrt{\sum_{j \in J} \kappa_j x_j}}, \tag{3.10}$$

and if $x_j = 0$, then we must have $\alpha_j^1 = 0$, which implies

$$\frac{R_j}{\kappa_j} \leq \frac{1}{2\sqrt{\sum_{j \in J} \kappa_j x_j}}. \tag{3.11}$$

Because the right-hand side of (3.11) is the same as that of (3.10), and this value is fixed at any point, this implies a strict ordering of R_j/κ_j values. (We assume these ratios are unique. If two or more demands exist with an identical ratio, we combine them into a single demand.) If we index demands based on this ratio, and if (3.10) holds for demand k, it must also hold for demand $k-1$. Therefore, if a KKT point exists such that $x_k = 1$, then a KKT point also exists with $x_{k-1} = 1$. The necessity of the KKT conditions implies the result.[1] □

This property implies that we can index demands in nonincreasing order of $R_j/D_j = r_j$, or unit revenue values, and an optimal solution will be of the form $x_j = 1$ for $j = 1, \ldots, k$ and $x_j = 0$ for $j = k+1, \ldots, n$, for some k between one and n. There are n solutions of this form, and we can sort demands and compute the average profit per unit time for each solution of this form in $\mathcal{O}(n \log n)$ time. If the solution with the maximum average profit per unit time among these n solution has positive average annual profit, then this solution is optimal; otherwise an optimal solution sets $x_j = 0$ for all $j \in J$ (and $Q = 0$).

3.1.2 The EPQ Problem with Market Choice

The standard EOQ model discussed in the previous section is sometimes referred to as having an infinite production rate, which is equivalent to the special case in which the delivery lead time equals zero. When the rate at which inventory is accumulated by the supplier is finite, the EOQ model must be generalized to account for this. Suppose that the supplier may only add to its inventory at a finite rate of P units per unit time. We will assume that this rate exceeds the maximum market demand rate (otherwise, accommodating this market without shortages would not be possible). Accounting for the effects of a finite production rate in the standard EOQ problem

[1] Note that this result can be shown by contradiction and using an interchange argument. That is, we suppose we have an optimal solution in which the property does not hold, and demonstrate that either a solution exists with the same objective function value satisfying the property, or that the initial solution is not optimal, a contradiction.

without market choice results in only a slight modification, which is often viewed as an adjustment to the holding cost by multiplying by a constant factor. The resulting model is often referred to as the economic production quantity (EPQ). Accounting for a finite production rate in the EOQ with market choice is a bit more involved, but leads to a similar result.

Under a finite production rate P, the average fixed order and holding costs per unit time for the standard EPQ model with demand rate D at the optimal order quantity may be written as

$$\sqrt{2SDH\left(1-\frac{D}{P}\right)}. \tag{3.12}$$

Using (3.12), along with market choice decision variables, the average profit function per unit time for the EPQ with market choice (EPQMC) problem is written as

$$\Pi^P(x)=\sum_{j\in J}R_jx_j-\sqrt{2SH\left(\sum_{j\in J}D_jx_j\right)\left(1-\frac{\sum_{j\in J}D_jx_j}{P}\right)}. \tag{3.13}$$

We can show that $\Pi^P(x)$ is convex in x (see [2]), and, as a result, a binary optimal solution exists when maximizing this function over $[0, 1]^n$, i.e., we can solve the continuous relaxation and obtain a binary optimal solution. As is shown in [2], the algorithm used to solve the EOQMC can be used to find an optimal solution for the EPQMC problem as well. We still index items in nonincreasing order of unit revenue (r_j) values and evaluate all solutions of the form $x_j = 1$ for $j = 1, \ldots, k$ and $x_j = 0$ for $j = k + 1, \ldots, n$, for some k between one and n. The only difference lies in the evaluation of the average profit per unit time for each solution vector, which uses (3.13) instead of (3.4).

3.2 EOQMC Problems with Constraints

Practical production and inventory planning problems must often incorporate practical constraints due to space, budget, or technological limits. We next consider the implications of certain types of constraints on the problems discussed in the previous section. These problems lead to interesting classes of knapsack problems with nonlinear objectives. We will first consider limits on the volume of demand that may be satisfied per unit time, followed by batch size constraints.

3.2.1 Demand Rate Constraints

Under a limit on the amount of demand that may be satisfied per unit time, we have a constraint of the form

$$\sum_{j\in J}D_jx_j\le b, \tag{3.14}$$

where b is the maximum demand rate that may be handled. While the nonlinear knapsack problem that results from maximizing either (3.4) or (3.13) over all $x \in \{\mathscr{B}^n \cap (3.14)\}$ is in general difficult (the recognition version in either case is $\mathscr{N}\mathscr{P}$-Complete), we can solve the continuous relaxations reasonably easily, as shown in [2].

For the EOQMC with the demand rate constraint, for example, an optimal solution to the continuous relaxation will exist at an extreme point, where extreme points correspond to either binary vectors that are feasible for (3.14), or solutions in which (3.14) is tight and at most one x_j variable is strictly between zero and one. The binary solutions generated in the algorithm for solving the unconstrained EOQMC that are feasible for (3.14) dominate all other binary vectors that are feasible for the constraint. Thus, we need to consider these solutions along with an optimal solution when (3.14) is tight in order to solve the continuous relaxation of the constrained problem. Fortunately, when (3.14) is tight, the square root terms in (3.4) and (3.13) become fixed constants. Thus, when the constraint is tight, the problem becomes a simple continuous linear knapsack problem, which is solved by inserting items into the knapsack in nonincreasing order of r_j values until the capacity is exhausted. Clearly this solution contains at most one fractional x_j, and a solution that is feasible for the binary restrictions can be obtained by simply setting the value of this fractional variable to zero. This rounding down heuristic is shown to be asymptotically optimal in the number of items in [2], under mild assumptions on the distributions of parameter values and the behavior of the capacity (and production rate, for the EPQMC) as $n \to \infty$. Because the continuous relaxation is relatively tight (and it has an optimal solution with at most one fractional variable), solution via customized branch-and-bound is generally quite effective.

3.2.2 Batch Size Constraints

In some contexts, a limit may exist on the maximum amount that may be produced or delivered in response to an order (due to, e.g., space limits, transportation capacity limits, or production technology restrictions). In this case, the associated constraint may be written very simply as

$$Q \leq b. \tag{3.15}$$

We illustrate the analysis of the problem under constraint (3.15) for the EOQMC case; for details on the EPQMC case, please see [2]. Recall that the unconstrained optimal order quantity is given in Eq. (3.2). However, given any binary vector x, it is possible to employ a batch size Q that is feasible for constraint (3.15), but does not take the form of an EOQ solution as in (3.2). We therefore consider two solution types. The first type of solution uses batches from the EOQ formula that are also feasible for (3.15). For these solutions, the batch size constraint may be written as

$$\sum_{j \in J} D_j x_j \leq \frac{b^2 H}{2S}, \tag{3.16}$$

which is exactly the form of constraint (3.14). Thus, we can use the methods described in the previous section for characterizing the best among solutions of this type.

The second type of solution is characterized by x vectors whose corresponding EOQMC batch size formula (3.2) is infeasible for (3.15). For such solution types, the convexity of the average costs per unit time in Q imply that constraint (3.15) must be tight, i.e., $Q = b$. When this is the case, the average holding plus fixed order costs per unit time as a function of the vector x, denoted as $AC(x)$, may be written as

$$AC(x) = \frac{S \sum_{j \in J} D_j x_j}{b} + \frac{Hb}{2}. \tag{3.17}$$

Finding the best solution of this type then requires solving the following problem:

$$\text{Maximize} \quad \sum_{j=1}^{n} \left(r_j - \frac{S}{b} \right) D_j x_j - \frac{Hb}{2} \tag{3.18}$$

$$\text{Subject to} \quad x_j \in \{0, 1\}, \quad j = 1, \ldots, n. \tag{3.19}$$

The above problem is easily solved by inspection by setting $x_j = 1$ for each demand j such that $r_j > S/b$ (interestingly, this solution may also be obtained exactly as before: by sorting demands in nonincreasing order of unit revenue values and evaluating all solutions of the form $x_j = 1$ for $j = 1, \ldots, k$ and $x_j = 0$ for $j = k+1, \ldots, n$, for some k between one and n). The resulting solution serves as the best solution of its type, and its average profit per unit time must then be compared with that of the best solution of the first type.

References

1. Bazaraa M, Sherali H, Shetty C (2006) Nonlinear Programming: Theory and Algorithms, 3rd edn. John Wiley & Sons, Hoboken, NJ
2. Geunes J, Shen Z, Romeijn H (2004) Economic Ordering Decisions with Market Choice Flexibility. Naval Research Logistics 51(1):117–136
3. Shen Z, Coullard C, Daskin M (2003) A Joint Location-Inventory Model. Transportation Science 37(1):40–55

Chapter 4
Single-Period Stochastic Inventory Planning with Demand Selection

Abstract This chapter deals with a generalization of the single-period newsvendor problem. We consider a setting in which a decision maker at a single stocking point must determine the stock level for a single product under uncertain demand. In addition to determining the item's stock level, the decision maker must select a subset from a set of individual demands, each of which is uncertain and follows a particular probability distribution. Assuming normally distributed and independent demand streams results in a class of problems that are strikingly similar to the problems considered in the previous chapter, although the underlying model assumptions are quite different.

4.1 The Selective Newsvendor Problem

We will first consider the basic version of what is referred to as the *selective newsvendor problem* (SNP), in which we must select a number of demand streams, along with a single stock level that will be used to satisfy these demand streams in a single-period setting. Following this, we consider more general contexts in which the decision maker can influence the distribution of each demand stream through marketing effort. Then we examine problems in which a limited marketing budget exists. Finally, we end the chapter by discussing a generalization of the problem in which price is a decision variable that directly influences the distribution parameters for each demand stream. The results in this chapter are based on more detailed analyses provided in [1] and [3].

4.2 The Basic SNP

Under the most basic version of the SNP, we consider a set J of n stochastic, single period demands, indexed by j. We assume throughout that each of these demands is normally distributed, that demand j has mean μ_j and standard deviation σ_j, and that all demand distributions are statistically independent. We assume that the product's unit procurement, holding and shortage costs are independent of the demand, although we permit demand-specific revenue values. That is, we define r_j as the

J. Geunes, *Demand Flexibility in Supply Chain Planning*,
SpringerBriefs in Optimization,
DOI 10.1007/978-1-4419-9347-2_4,

per-unit net revenue for demand j (in excess of the unit cost C). Letting x_j denote a binary variable equal to one if we select demand j (and zero otherwise), then the aggregated demand for the product is normally distributed with mean $\sum_{j \in J} \mu_j x_j$ and standard deviation $\sqrt{\sum_{j \in J} \sigma_j^2 x_j}$. Using the single-period inventory model defined in Eq. (1.10), and replacing the per-unit net revenue term $r = p - C$ with the demand-specific net revenue associated with each demand r_j, we can write the single-period expected profit for the selective newsvendor as

$$\Pi_n(Q^*, x) = \sum_{j \in J} r_j \mu_j x_j - K(z^*) \sqrt{\sum_{j \in J} \sigma_j^2 x_j}. \tag{4.1}$$

Somewhat surprisingly, although the SNP model arises from a different set of modeling assumptions, Eq. (4.1) has precisely the same mathematical form as the average profit per unit time equation for the EOQMC (see Eq. (3.4)). Property 3.1 immediately implies that we can maximize $\Pi_n(Q^*, x)$ over all $x \in \mathscr{B}^n$ by indexing demands in nonincreasing order of the ratio $r_j \mu_j / \sigma_j^2$, and evaluating the expected profit for each of the n solutions of the form $x_j = 1$ for $j = 1, \ldots, k$ and $x_j = 0$ for $j = k+1, \ldots, n$, for $k = 1, \ldots, n$. If the solution with the maximum expected profit among these n solutions has positive expected profit, then this solution is optimal; otherwise an optimal solution sets $x_j = 0$ for all $j \in J$ (and $Q = 0$).

4.3 The SNP with Market Effort

The SNP model in the previous section assumes that the distribution of each demand has fixed parameters. We now consider situations in which the supplier can influence these parameters. For example, if each demand j corresponds to a market (or market segment), then we permit the supplier to exert some effort to affect the market's demand distribution. This effort might, for example, come in the form of advertising expenditures. We assume that the amount of market effort exerted in market j, which we denote as a_j, can be measured in discrete units, and that the cost per unit of effort equals t_j. We also assume that the expected demand (demand variance) in market j can be expressed as a function of this market effort, and we express this function as $\mu_j(a_j)$ ($\sigma_j^2(a_j)$). These functions are assumed to be continuous, nonnegative, nondecreasing in a_j, and bounded. Further, we assume that a value δ_j (Δ_j) exists such that for all $a_j \geq \delta_j$ ($a_j \geq \Delta_j$), $\mu_j(a_j) = \bar{\mu}_j$ ($\sigma_j^2(a_j) = \bar{\sigma}_j^2$), i.e., a threshold value exists such that these functions are constant (or "level off") when this threshold is exceeded. We further generalize the SNP model in this section to permit a fixed cost for entry into market j, i.e., let S_j denote the fixed cost for selling in market j. The following subsection considers the case in which market variance is independent of market effort (i.e., $\sigma_j^2(a_j) = \sigma_j^2$ for all $a_j \geq 0$), followed by the case in which market variance may depend on market effort.

4.3.1 Market Variance Independent of Market Effort

When each market's variance is independent of market effort, we need to first determine the optimal effort exerted in market j, if we choose market j. This is accomplished by determining the value of a_j that maximizes $\theta_j(a_j) = r_j\mu_j(a_j) - t_j a_j$. The difficulty of determining this value of a_j depends on the functional form of $\mu_j(a_j)$. If, for example, each function $\theta_j(a_j)$ is differentiable and concave, then a simple first-order condition is all that is required for identifying an optimal value of a_j, denoted by a_j^*. This first-order condition can be written as

$$\frac{d\mu_j(a_j^*)}{da_j} = \frac{t_j}{r_j}. \tag{4.2}$$

Letting $R_j^* = r_j\mu_j(a_j^*) - t_j a_j^* - S_j$, then we can determine an optimal solution for the case in which expected demand depends on market effort by maximizing

$$\Pi_n(Q^*, x) = \sum_{j \in J} R_j^* x_j - K(z^*)\sqrt{\sum\nolimits_{j \in J} \sigma_j^2 x_j}, \tag{4.3}$$

over all $x \in \mathscr{B}^n$. Clearly (4.3) takes the same form as (4.1), and we can solve this problem using the sorting approach we have described based on the ratios R_j^*/σ_j^2 for $j = 1, \ldots, n$.

4.3.2 Market Variance Dependent on Market Effort

When each market's demand variance also depends on the level of market effort, then the functional form of each $\mu_j(a_j)$ and $\sigma_j^2(a_j)$ function determines the problem's overall complexity. In [3], specific functional forms are assumed, which may be characterized as follows. Each expected demand function $\mu_j(a_j)$ is an approximate S-curve function that is nonnegative and continuous for all $a_j \geq 0$, strictly convex and increasing in a_j for $0 \leq a_j \leq \alpha_j$, linearly increasing for $\alpha_j \leq a_j \leq \delta_j$ (with a slope at least as great as the limit of the derivative of the strictly convex portion as $a_j \to \alpha_j$), and fixed at $\bar{\mu}_j$ for $a_j \geq \delta_j$. The variance function is nonnegative, continuous, concave, and nondecreasing for all $a_j \geq 0$, with $\sigma_j^2(a_j) = \bar{\sigma}_j^2$ for all $a_j \geq \Delta_j$ for some positive Δ_j. Under these assumptions, it is shown in [3] that the optimal effort level in market j, if market j is selected, must equal either 0 or δ_j. Assuming that the fixed cost of market entry, S_j, is at least as great as the expected net revenue at zero effort (at $a_j = 0$), then we can set $R_j^* = r_j\mu_j(\delta_j) - t_j\delta_j - S_j$ (which is equivalent to setting $a_j^* = \delta_j$ in our approach in the previous subsection), and apply the same sorting-based solution approach as before, again using the ratio R_j^*/σ_j^2.

4.4 The SNP with Limited Market Resources

In most situations, resources required for marketing effort are not unlimited. We therefore assume that the total amount of market effort is limited by some value b. For ease of exposition, we assume that $\mu_j(0) = 0$, $\sigma_j^2(0) = 0$, and $S_j > 0$ for all $j \in J$.[1] Note that this implies $S_j > r_j\mu_j(0)$ for all $j \in J$. We also assume that $\mu_j(a_j)$ is convex and that $\sigma_j^2(a_j)$ is concave, in addition to the previous assumptions we have stated regarding these functions. We formulate the SNP with limited market resources as follows:

$$\text{[SNPM]} \quad \text{Maximize} \quad \sum_{j\in J}(R_j(a_j) - t_j a_j - S_j)x_j - K(z^*)\sqrt{\sum_{j\in J}\sigma_j^2(a_j)x_j} \tag{4.4}$$

$$\text{Subject to} \quad \sum_{j\in J} a_j \le b, \tag{4.5}$$

$$x_j \in \{0,1\}, \quad \forall j \in J, \tag{4.6}$$

where $R_j(a_j) = r_j\mu_j(a_j)$. The SNPM formulation is quite complicated for several reasons, not the least of which is the presence of the product of continuous functions of continuous variables and binary variables. Unlike the approach in the previous section, the a_j values are no longer independent, and we cannot, therefore, simply determine the optimal value of a_j by maximizing $\theta_j(a_j)$ over $a_j \ge 0$. Our goal is therefore to provide an effective method for efficient solution of the continuous relaxation of SNPM, which can then be used within a branch-and-bound scheme.

To this end, observe that $a_j \le \delta_j x_j$ is a valid inequality for SNPM for all $j \in J$, as $\mu_j(a_j) = \bar{\mu}_j$ for $j \ge \delta_j$. We next define a new set of continuous variables w_j for $j \in J$, such that $w_j = a_j/\delta_j$, which implies that we can confine our attention to continuous values of w_j between zero and one, and we can rewrite the SNPM as

$$\text{Maximize} \quad \sum_{j\in J}(R_j(\delta_j w_j) - t_j\delta_j w_j - S_j x_j) - K(z^*)\sqrt{\sum_{j\in J}\sigma_j^2(\delta_j w_j)} \tag{4.7}$$

$$\sum_{j\in J}\delta_j w_j \le b, \tag{4.8}$$

$$x_j \in \{0,1\}, \quad \forall j \in J, \tag{4.9}$$

$$0 \le w_j \le x_j, \quad \forall j \in J. \tag{4.10}$$

We next observe that if we do not select market j, which implies $x_j = 0$, then an optimal solution exists such that $w_j = 0$, i.e., we can change the right-hand

[1] For a detailed discussion of the more general case in which $\mu_j(0) > 0$ and $\sigma_j^2(0) > 0$, please see [3].

side of each constraint in constraint set (4.10) to one without loss of optimality. We can actually demonstrate a stronger result for the continuous relaxation (see [3]). That is, when we relax (4.9) to permit $x_j \in [0, 1]$ for each $j \in J$, then we can show that an optimal solution exists for this relaxation such that $w_j = x_j$, which is equivalent to $a_j = \delta_j x_j$. Next, note that because $\mu_j(a_j)$ is convex, we have $\mu_j(\delta_j x_j) \leq \mu_j(\delta_j)x_j$; similarly, because $\sigma_j^2(a_j)$ is concave we also have $\sigma_j^2(\delta_j)x_j \leq \sigma_j^2(\delta_j x_j)$.

As a result, we obtain the following formulation for solving the continuous relaxation of SNPM:

$$\text{Maximize} \quad \sum_{j \in J} \tilde{R}_j(\delta_j)w_j - K(z^*)\sqrt{\sum_{j \in J} \sigma_j^2(\delta_j)w_j} \tag{4.11}$$

$$\text{Subject to} \quad \sum_{j \in J} \delta_j w_j \leq b, \tag{4.12}$$

$$0 \leq w_j \leq 1, \quad \forall j \in J, \tag{4.13}$$

where $\tilde{R}_j(\delta_j) = R_j(\delta_j) - t_j\delta_j - S_j$. The objective function of the above problem is precisely the same as that of the SNP and EOQMC problems. However, the knapsack constraint (4.12) complicates the problem and leads to an interesting class of nonlinear knapsack problems. Unfortunately, this problem is not as nicely structured as the nonlinear knapsack problems considered in the previous chapter. Despite this, it is possible to solve this relaxation in $\mathcal{O}(n^3)$ time, as shown in [2]. For a detailed discussion of a customized branch-and-bound algorithm for solving SNPM, please see [3].

4.5 The SNP with Pricing

So far in this chapter, we have not considered the impact of pricing on the solution of the SNP. We next consider a profit-maximizing model in which prices are decision variables. We first consider the case in which no price discrimination is possible, i.e., the price must be the same for every market. We then discuss the problem when price discrimination is permitted.

4.5.1 Equal Market Prices

The SNP with equal market prices, which we denote by SNPP$_=$, can be formulated as follows:

$$[\text{SNPP}_=] \quad \text{Maximize} \quad \sum_{j \in J} R_j(p)x_j - K(z^*)\sqrt{\sum_{j \in J} \sigma_j^2(p)x_j} \tag{4.14}$$

$$\text{Subject to} \quad x_j \in \{0, 1\}, \quad \forall j \in J, \tag{4.15}$$

where $R_j(p)$ denotes the expected net revenue in market j as a function of price p, and $\sigma_j^2(p)$ denotes the variance in market j as a function of p. We assume that for each market, some maximum price level exists such that for prices exceeding this level, demand in the market equals zero. Let p_j^0 denote the price in market j such that for $p \geq p_j^0$, demand in market j equals zero. Note that for any fixed price, p, the above problem becomes the basic SNP, and can thus be solved efficiently based on our ratio sorting scheme, using the ratio $R_j(p)/\sigma_j^2(p)$. When the price varies, however, the ratio ordering of the markets may change. If we can characterize a manageable set of price intervals such that the ratio ordering within an interval does not change, then we can solve an SNP problem for each interval and determine the best price for that interval.

Let P_{ij} denote the critical price value at which the preference ratios for markets i and j are equal, i.e.,

$$P_{ij} = \left\{C \leq p \leq \min\{p_i^0, p_j^0\} : \frac{R_i(p)}{\sigma_i^2(p)} = \frac{R_j(p)}{\sigma_j^2(p)}\right\}, \tag{4.16}$$

where we assume that the price must be greater than or equal to the unit variable cost C. We assume that a finite number of such critical price levels exists according to (4.16) (the number of prices satisfying (4.16) depends on the functional forms of the revenue and variance functions, but this number is often one for simple forms of these functions; if this is the case, then the total number of such critical price levels is $\mathcal{O}(n^2)$). Given these values, we create an ordered list of price values, denoted by p^k for $k = 1, \ldots, V$, containing each P_{ij} value, plus all p_j^0 values. This sequence is defined as $c = p^0 < p^1 < \cdots < p^V$. Within an interval (p^k, p^{k+1}), the ratio ordering for all markets does not change. Moreover, for consecutive intervals (p^{k-1}, p^k) and (p^k, p^{k+1}), the rank order of markets will be the same except for markets i and j, when $p^k \in P_{ij}$. This implies that we only need to rank order the markets once, and then switch the order of the appropriate markets when moving from one interval to the next. Within each price interval (p^{k-1}, p^k), we solve n optimization problems of the form

$$\text{Maximize}_{p \in (p^{k-1}, p^k)} \left\{\sum_{j=1}^{l} R_j(p) - K(z^*)\sqrt{\sum_{j=1}^{l} \sigma_j^2(p)}\right\}, \tag{4.17}$$

for $l = 1, \ldots, n$. If one value exists that satisfies (4.16) for each pair of markets[2], then the total time required to determine an optimal solution is $\mathcal{O}(Tn^3)$, where we assume that the time required to solve (4.17) is $\mathcal{O}(T)$.

[2] Please see [1] for a discussion of fairly general models for which this property holds, i.e., at most one value exists satisfying (4.16) for each pair of markets.

4.5.2 SNP with Market Price Discrimination

We next consider the case in which different prices may be set for individual markets. For this case, we let p_j denote the price in market j, and we assume that a price p_j^0 exists such that for all $p_j \geq p_j^0$, both the revenue function $R_j(p_j)$ and the variance function $\sigma_j^2(p_j)$ equal zero. This allows formulating the problem without binary selection variables, since we can force both zero revenue and zero cost in a market by setting a sufficiently high price for that market. We can therefore formulate the problem with individual market prices as follows:

$$[\text{SNPP}_{\neq}] \quad \text{Maximize} \quad \sum_{j\in J} R_j(p_j) - K(z^*)\sqrt{\sum_{j\in J} \sigma_j^2(p_j)} \tag{4.18}$$

$$\text{Subject to} \quad 0 \leq p_j \leq p_j^0, \quad \forall j \in J. \tag{4.19}$$

If the revenue function is concave in price and the variance function is convex in price, then this problem is a convex program. In this case, we can show that an optimal solution exists such that either all markets are "selected" or none are selected (see [1]; of course, a market may be selected in the sense that an arbitrarily small but positive demand level is satisfied).

References

1. Bakal I, Geunes J, Romeijn H (2008) Market Selection Decisions for Inventory Models with Price-Sensitive Demand. Journal of Global Optimization 41(4):633–657
2. Romeijn H, Geunes J, Taaffe K (2007) Solution Methods for a Class of Nonlinear, Non-Separable Knapsack Problems. Operations Research Letters 35(2):172–180
3. Taaffe K, Geunes J, Romeijn H (2008) Target Market Selection with Demand Uncertainty: The Selective Newsvendor Problem. European Journal of Operational Research 189(3):987–1003

Chapter 5
Dynamic Lot Sizing with Demand Selection and the Pricing Analog

Abstract This chapter defines a model for production planning for a single product in a periodic setting, where the planner must select from a number of individual orders for the product. Associated with any order are a demand quantity, delivery period, and revenue. Acceptance of an order implies that it must be met on time and in full. Orders are placed in advance of the planning horizon, and the planner must determine which orders to accept as well as a production plan for meeting these orders on time over a finite horizon. Production in any period carries a fixed order cost as well as variable production costs, and inventory may be held from period to period, incurring an associated holding cost. The planner's goal is to maximize profit from order acceptance decisions over the planning horizon.

5.1 Demand Selection Problem Definition

The model we consider in this chapter generalizes the ELSP, discussed in Chap. 1, and is intimately related to the ELSP with pricing presented in Chap. 2, as we later discuss in Sect. 5.4. In this section we formally define the *demand selection problem* (DSP), while the remainder of this chapter provides a solution algorithm (Sect. 5.2), discusses the implications of capacity limits on production (Sect. 5.5), and provides an interpretation of the problem as an equivalent pricing and lot sizing problem (Sect. 5.4). Much of the work presented in this chapter is based on [3].

Consider a single-item production planning problem for a finite planning horizon of length T, where production occurs and demand is realized in a discrete set of periods indexed by t. In advance of the planning horizon, a number of demands are available, and each of these demands must be either accepted or rejected. Associated with demand j in period t are a demand quantity, d_{jt}, and a per-unit revenue, r_{jt}; thus, the total revenue associated with acceptance of demand j in period t equals $r_{jt}d_{jt}$. We assume without loss of generality that n_t demands exist in period t, and that demands within any period are indexed in non-increasing order of unit revenues, i.e., $r_{jt} \geq r_{j+1,t}$, for all $t = 1, \ldots, T$ and $j = 1, \ldots, n_t - 1$.

Let w_{jt} denote the percentage of demand j in period t that is fulfilled, i.e., w_{jt} is a continuous variable between zero and one (as we later show, an optimal solution will exist in which each w_{jt} is strictly zero or one in the absence of capacity limits). The total demand that must be met in period t, which depends on the demand

J. Geunes, *Demand Flexibility in Supply Chain Planning*,
SpringerBriefs in Optimization,
DOI 10.1007/978-1-4419-9347-2_5,

acceptance decisions, equals $\sum_{j=1}^{n_t} d_{jt} w_{jt}$. Using the notation and parameter and decision variable definitions used for the ELSP, we formulate the DSP as follows:

$$\text{[DSP]} \quad \text{Maximize} \quad \sum_{t=1}^{T} \left\{ \sum_{j=1}^{n_t} r_{jt} d_{jt} w_{jt} - S_t y_t - C_t Q_t - H_t I_t \right\} \tag{5.1}$$

$$\text{Subject to} \quad I_t = Q_t + I_{t-1} - \sum_{j=1}^{n_t} d_{jt} w_{jt}, \quad t = 1, \ldots, T, \tag{5.2}$$

$$Q_t \le M_t y_t, \quad t = 1, \ldots, T, \tag{5.3}$$

$$I_0 = 0, \qquad Q_t, I_t \ge 0, \quad t = 1, \ldots, T, \tag{5.4}$$

$$0 \le w_{jt} \le 1, \quad t = 1, \ldots, T, \quad j = 1, \ldots, n_t \tag{5.5}$$

$$y_t \in \{0, 1\}, \quad t = 1, \ldots, T. \tag{5.6}$$

The DSP formulation is based on the standard formulation of the ELSP. This formulation is not the tightest possible formulation, i.e., it is possible to reformulate the problem such that the linear programming (LP) relaxation value provides a better upper bound on the optimal solution to the DSP. This reformulation of the LP relaxation is closely related to the facility location reformulation of ELSP discussed in Sect. 1.2.6. We begin by defining the modified revenue parameter $\rho_{jt} = d_{jt}(r_{jt} + \sum_{\tau=t}^{T} H_t)$ as well as the modified cost parameter $\hat{C}_t = C_t + \sum_{\tau=t}^{T} H_t$. Substituting $I_t = \sum_{\tau=1}^{t} Q_\tau - \sum_{\tau=1}^{t} \sum_{j=1}^{n_t} d_{jt} w_{jt}$ and letting $Q_{jt\tau}$ denote the number of units produced in period t to satisfy demand j in period τ, we can reformulate the DSP as follows:

$$\text{[FDSP]} \quad \text{Minimize} \quad \sum_{t=1}^{T} \left\{ S_t y_t + \hat{C}_t \sum_{\tau=t}^{T} \sum_{j=1}^{n_t} Q_{jt\tau} - \sum_{j=1}^{n_t} \rho_{jt} w_{jt} \right\} \tag{5.7}$$

$$\text{Subject to} \quad \sum_{t=1}^{\tau} Q_{jt\tau} = d_{j\tau} w_{j\tau}, \quad \tau = 1, \ldots, T, \quad j = 1, \ldots, n_\tau, \tag{5.8}$$

$$\sum_{j=1}^{n_t} Q_{jt\tau} \le M_{t\tau} y_t, \quad t = 1, \ldots, T, \quad \tau = t, \ldots, T, \tag{5.9}$$

$$0 \le w_{j\tau} \le 1, \quad \tau = 1, \ldots, T, \quad j = 1, \ldots, n_\tau \tag{5.10}$$

$$y_t, Q_{jt\tau} \ge 0, \quad t = 1, \ldots, T, \quad \tau = t, \ldots, T, \quad j = 1, \ldots, n_t. \tag{5.11}$$

The multiplier on the right-hand side of constraint set (5.9), $M_{t\tau}$, is a large number that ensures the problem remains uncapacitated, and can be replaced with $M_{t\tau} = \sum_{j=1}^{n_\tau} d_{j\tau}$ without loss of optimality. To demonstrate the tightness of the LP relaxation FDSP, the next section provides an algorithm for solving the dual of FDSP. We then show that the resulting dual has an optimal solution with a complementary primal solution that is feasible for DSP, and therefore optimal. We note that this reformulation and dual solution approach follow that shown in [6].

5.2 Dual Ascent Solution Algorithm

This section formulates the dual problem for FDSP and provides an algorithm for solving this dual problem. We first define dual multipliers $\mu_{j\tau}$, $\omega_{t\tau}$, and $\pi_{j\tau}$ associated with primal constraints (5.8), (5.9), and the upper bounding constraints of (5.10). We can then formulate the dual problem, DP, as follows:

$$\text{[DP]} \quad \text{Maximize} \quad \sum_{\tau=1}^{T}\sum_{j=1}^{n_t} -\pi_{j\tau} \tag{5.12}$$

$$\text{Subject to} \quad \sum_{\tau=t}^{T}\sum_{j=1}^{n_t} d_{j\tau}\omega_{t\tau} \leq S_t, \qquad t = 1, \ldots, T, \tag{5.13}$$

$$d_{j\tau}\mu_{j\tau} + \pi_{j\tau} \geq \rho_{j\tau}, \qquad \tau = 1, \ldots, T, j = 1, \ldots, n_\tau \tag{5.14}$$

$$\mu_{j\tau} - \omega_{t\tau} \leq \hat{C}_t, \qquad t = 1, \ldots, T, \tau = t, \ldots, T, \; j = 1, \ldots, n_t, \tag{5.15}$$

$$\pi_{j\tau}, \omega_{t\tau} \geq 0, \mu_{j\tau} \text{ unrestricted}, \qquad t = 1, \ldots, T, \tau = t, \ldots, T, \; j = 1, \ldots, n_t. \tag{5.16}$$

We observe from (5.15) that an optimal solution will exist such that $\omega_{t\tau}$ is equal to the maximum between zero and $\mu_{j\tau} - \hat{C}_t$ for all t, τ, and j. In addition, because of the form of the objective and constraint set (5.14), at optimality, we will have $-\pi_{j\tau}$ equal to the minimum between zero and $d_{j\tau}\mu_{j\tau} - \rho_{j\tau}$ for every j and τ. Based on these observations, we can formulate DP much more compactly as follows:

$$\text{[CDP]} \quad \text{Maximize} \quad \sum_{\tau=1}^{T}\sum_{j=1}^{n_t} \min\{0, d_{j\tau}\mu_{j\tau} - \rho_{j\tau}\} \tag{5.17}$$

$$\text{Subject to} \quad \sum_{\tau=t}^{T}\sum_{j=1}^{n_t} d_{j\tau}\left(\max\{0, \mu_{j\tau} - \hat{C}_t\}\right) \leq S_t, \quad t = 1, \ldots, T. \tag{5.18}$$

Inspection of CDP leads to some interesting and useful observations. Note first that an optimal solution exists such that $\mu_{j\tau}$ is no greater than $\rho_{j\tau}/d_{j\tau}$ for every j, τ pair. Thus, for any j, τ pair such that $\min_{t=1,\ldots,\tau} \hat{C}_t \geq \rho_{j\tau}/d_{j\tau}$, the variable $\mu_{j\tau}$ can be set to its maximum value of $\rho_{j\tau}/d_{j\tau}$ without consuming any constraint "capacity," which implies that we can eliminate such variables from the formulation immediately. Next, note that we can start with an initial feasible solution that sets $\mu_{j\tau} = \min_{t=1,\ldots,\tau} \hat{C}_t$ for every j, τ pair. This initial solution does not consume any of the capacity of any of the constraints (5.18). Beginning with this initial solution, we will increase values of the $\mu_{j\tau}$ variables in a specific order, resulting in what is commonly called a dual ascent algorithm.

This algorithm begins by simultaneously increasing the values of all μ_{j1} variables from their initial value of $\hat{C}_1$. That is, we create a set J_1^1 containing all orders in period 1 that have not been eliminated. Increasing these variables consumes some of the capacity, S_1, of constraint 1, but does not affect any of the other constraints, as the μ_{j1} variables appear only in the first constraint. If μ_{k1} hits the value ρ_{k1}/d_{k1} before constraint 1 becomes tight, then we remove order k from the set J_1^1, insert order k into the set J_1^0, and we do not continue to increase the value of μ_{k1} with the other μ_{j1} variables such that $j \in J_1^1$. We continue simultaneously increasing the μ_{j1} values for all $j \in J_1^1$, until either $J_1^1 = \emptyset$ or until the variables consume all of the capacity of constraint 1. We therefore have

$$\mu_{j1} = \begin{cases} \frac{\rho_{j1}}{d_{j1}}, & j \in J_1^0, \\ \hat{C}_1 + \frac{S_1 - \sum_{j \in J_1^0} d_{j1} \max\{0, \rho_{j1}/d_{j1} - \hat{C}_1\}}{\sum_{j \in J_1^1} d_{j1}}, & j \in J_1^1. \end{cases} \tag{5.19}$$

We next move on to period 2, by inserting all orders in period 2 in the set J_2^1, and simultaneously increasing the values of all μ_{j2} variables for $j \in J_2^1$. A variable μ_{k2} may be blocked from further increase because (a) it hits the value ρ_{k2}/d_{k2}, in which case order k is removed from J_2^1 and inserted in the set J_2^0; (b) constraint 2 becomes tight; or (c) it hits the value $\hat{C}_1$ (and any further increase would violate constraint 1). At this point we have

$$\mu_{j2} = \begin{cases} \frac{\rho_{j2}}{d_{j2}}, & j \in J_2^0, \\ \min\{\hat{C}_1; \hat{C}_2 + \frac{S_2 - \sum_{j \in J_2^0} d_{j2} \max\{0, \frac{\rho_{j2}}{d_{j2}} - \hat{C}_2\}}{\sum_{j \in J_2^1} d_{j2}}\}, & j \in J_2^1. \end{cases} \tag{5.20}$$

We continue this approach for subsequent periods. In general, for period t, if $J_t^1 \neq \emptyset$, then letting $[x]^+ = \max\{0, x\}$, we can write the value of μ_{kt} for $k \in J_t^1$ using the formula

$$\mu_t^* = \min_{s \leq t} \left\{ \hat{C}_s + \frac{S_s - \sum_{\tau=s}^{t} \sum_{j \in J_t^0} d_{j\tau} [\frac{\rho_{j\tau}}{d_{j\tau}} - \hat{C}_s]^+ - \sum_{\tau=s}^{t-1} \sum_{j \in J_\tau^1} d_{j\tau} [\mu_\tau^* - \hat{C}_s]^+}{\sum_{j \in J_t^1} d_{jt}} \right\}, \tag{5.21}$$

which implies that we can write

$$\mu_{jt} = \begin{cases} \frac{\rho_{jt}}{d_{jt}}, & j \in J_t^0, \\ \mu_t^*, & j \in J_t^1, \end{cases} \tag{5.22}$$

for $t = 1, \ldots, T$ and $j = 1, \ldots, n_t$. We can show that the dual solution above, obtained via the dual ascent approach, solves DP (see [2]). We can obtain a complementary primal solution as follows. If constraint t is tight in an optimal solution, then this implies that $y_t = 1$ and an order is placed in period t; otherwise $y_t = 0$. For order–period pair (j, t), if $j \in J_t^0$, then $w_{jt} = 0$, and demand j in period t is not selected. If $j \in J_t^1$, then demand j in period t is selected and $w_{jt} = 1$. Moreover, for any demand $j \in J_t^1$, $\mu_{jt} = \mu_t^*$, and production for the demand occurs in the period s that gives the minimum in Eq. (5.21). This implies that for $j \in J_t^1$, if $s^*(t)$ denotes the index s that provides the minimum in (5.21), then $Q_{js^*(t)t} = d_{jt}$, and $Q_{jst} = 0$ for all $s \neq s^*(t)$. Moreover, for all pairs (j, t) such that $j \in J_t^0$, $Q_{jst} = 0$ for all $s = 1, \ldots, T$. This solution is clearly feasible for the DSP (including the binary conditions), and we can show that its objective function value is the same as the corresponding complementary dual solution objective function value for DP (see [2]). The resulting solution is therefore optimal for the DSP, and the formulation FDSP therefore has zero integrality gap. If we let $n_{\max} = \max_{t=1,\ldots,T}\{n_t\}$, then we can express the worst-case solution time of this dual ascent approach as $\mathcal{O}(n_{\max} T^2)$.

5.3 Shortest Path Solution Approach

This section describes a more direct approach for solving the DSP, using the shortest path[1] solution structure for the ELSP described in Sect. 1.2.3. Like the lot sizing with pricing problem discussed in Sect. 2.3, for a given set of selected orders, the DSP is equivalent to the ELSP. Recall that the shortest path graph structure for solving an instance of the ELSP contains $T + 1$ nodes, with an arc from node t to node $s + 1$ if $s \geq t$ for all $t = 1, \ldots, T$. Arc $(t, s + 1)$ implies that production in period t is used to satisfy demands in periods t through s. For the DSP, we create the same graph structure, but we must solve a subproblem in order to determine which demands should be selected if arc $(t, s + 1)$ is used, for $t = 1, \ldots, T$ and $s = t, \ldots, T$. This allows us to associate a net profit level with each arc, and subsequently solve the longest path problem on the resulting acyclic graph.

Solving the arc subproblem to determine the maximum profit associated with the arc $(t, s + 1)$ requires solving the following problem, where we again define $H_{t,\tau} = \sum_{u=t}^{\tau-1} H_u$:

$$\text{Maximize} \quad \phi(t, s + 1) = \sum_{\tau=t}^{s} \sum_{j=1}^{n_\tau} (r_{j\tau} - C_t - H_{t,\tau}) d_{j\tau} w_{j\tau} \tag{5.23}$$

[1] We actually solve a longest-path problem on an acyclic graph, although we use the shortest-path terminology due to its prominent usage in the literature.

$$\text{Subject to} \quad w_{j\tau} \in \{0, 1\}, \quad \tau = t, \ldots, s, \; j = 1, \ldots, n_\tau. \tag{5.24}$$

For the given arc $(t, s+1)$, clearly we will select demand j in period τ such that $t \le \tau \le s$ if $r_{j\tau} > C_t + H_{t,\tau}$, i.e., if the unit revenue exceeds the variable costs associated with the demand. If the optimal solution to the above subproblem, which we denote as $\phi^*(t, s+1)$, exceeds the fixed order cost S_t, then the arc is assigned a net profit of $\phi^*(t, s+1) - S_t$. Otherwise, the arc is assigned a net profit of zero.

This solution approach is much easier to describe than the dual ascent method, although its complexity is the same, equal to $\mathcal{O}(n_{\max} T^2)$. The dual ascent method, however, enables demonstrating that the FDSP formulation has zero integrality gap, i.e., that this formulation is tight.

5.4 Interpretation of the DSP as a Pricing Problem

Recall that the ELSP with pricing problem described in Sect. 2.3 assumed that demand in each period was price dependent, using the function $D_t(p_t)$ to express demand in period t as a function of the price in period t, p_t. The total revenue in this model, contained in the first term of the objective function of [ELSP′], was equal to $p_t D_t(p_t)$. For many common forms of the demand function, $D_t(p_t)$, the resulting *revenue function*, equal to $p_t D_t(p_t)$, is a concave function (for example, when the demand function is linear in price). If we have a one-to-one correspondence between the price in period t and the demand in period t, then, without loss of generality, we can express the revenue function in a period as a function of either the price or the demand in period t. If we choose the latter, then we can write the revenue function in period t as the function $R_t(D_t)$. That is, replacing $D_t(p_t)$ with D_t and replacing $p_t D_t(p_t)$ with $R_t(D_t)$ in the formulation ELSP′ in Sect. 2.3 results in an equivalent formulation in the demand variables. This reformulation may be expressed as follows:

$$[\text{ELSP}'(\text{D})] \quad \text{Maximize} \quad \sum_{t=1}^{T} \{R_t(D_t) - S_t y_t - C_t Q_t - H_t I_t\} \tag{5.25}$$

$$\text{Subject to} \quad I_t = Q_t + I_{t-1} - D_t, \quad t = 1, \ldots, T, \tag{5.26}$$

$$Q_t \le M_t y_t, \quad t = 1, \ldots, T, \tag{5.27}$$

$$Q_t, I_t, D_t \ge 0, \quad t = 1, \ldots, T, \tag{5.28}$$

$$y_t \in \{0, 1\}, \quad t = 1, \ldots, T. \tag{5.29}$$

Suppose that we can approximate the revenue function in any period t as a piecewise linear, concave, and nondecreasing function of the demand level in period t, with $n_t + 1$ segments. Letting r_{jt} denote the slope of segment j of the revenue function in period t, and assuming $r_{jt} > r_{j+1,t}$, then we can express this revenue function as

$$R_t(D_t) = \begin{cases} \sum_{j=1}^{k-1} r_{jt} d_{jt} + r_{kt}(D_t - \sum_{j=1}^{k-1} d_{jt}), & \sum_{j=1}^{k-1} d_{jt} \le D_t < \sum_{j=1}^{k} d_{jt}, \\ & k = 1, \dots, n_t, \\ \sum_{j=1}^{n_t} r_{jt} d_{jt}, & D_t \ge \sum_{j=1}^{n_t} d_{jt}. \end{cases} \tag{5.30}$$

Using this form of the revenue function in the lot sizing and pricing problem ELSP′(D), then, for the subproblem corresponding to arc $(t, s+1)$, according to Eq. (2.15), we need to solve

$$\text{[PSP(D)]} \quad \text{Maximize} \quad \sum_{\tau=t}^{s} \left\{ R_\tau(D_\tau) - (C_t + H_{t,\tau}) D_\tau \right\}. \tag{5.31}$$

To model the revenue functions shown in Eq. (5.30), we create a binary variable $w_{j\tau}$ associated with the jth linear segment of the revenue function in period τ, for $\tau = 1, \dots, T$, and $j = 1, \dots, n_\tau + 1$. The resulting arc profit subproblem takes the exact form of (5.23)–(5.24) (with an additional zero-slope segment for every period). Observe that because we are maximizing, we need not impose explicit constraints on the selection of different linear segments associated with the revenue function in any given period, i.e., an optimal solution will select segments in nonincreasing order of segment slopes. Thus, this special case of the ELSP with pricing and piecewise linear and concave revenue functions is equivalent to the DSP.

Before proceeding we note two additional properties of the arc subproblem [PSP(D)] (which also hold for the arc subproblem expressed in (5.23)–(5.24), as these problems are equivalent) for any arc $(t, s+1)$. First, note that explicit binary restrictions on the $w_{j\tau}$ variables are not required, as an optimal solution will exist for the continuous relaxation such that each $w_{j\tau}$ takes a value of zero or one. In particular, an optimal solution exists such that $w_{j\tau} = 1$ for every pair (j, τ) such that $r_{j\tau} > C_t + H_{t,\tau}$, and $w_{j\tau} = 0$ for every (j, τ) such that $r_{j\tau} \le C_t + H_{t,\tau}$. This leads to the second property, which states that for arc $(t, s+1)$ and for period τ such that $t \le \tau \le s$, the generalized first-order condition for (5.31) is necessary and sufficient for optimality, i.e., $C_t + H_{t,\tau} \in \partial R_\tau(D_\tau)$, where $\partial R_\tau(D_\tau)$ is the set of subgradients of the function $R_\tau(D_\tau)$ at D_τ. However, this condition is equivalent to the condition that if $r_{(k+1)\tau} \le C_t + H_{t,\tau} \le r_{k\tau}$, then $C_t + H_{t,\tau} \in \partial R_\tau(\sum_{j=1}^{k} d_{j\tau})$, and an optimal solution exists for the arc $(t, s+1)$ subproblem such that $w_{j\tau} = 1$ for $j = 1, \dots, k$ and $w_{j\tau} = 0$ for $j = k+1, \dots, n_\tau + 1$.

5.5 Capacitated Versions

Our analysis in this chapter thus far has assumed that no limit exists on the size of any production/procurement order. In contexts in which production/procurement orders are limited by some capacity level (which includes virtually all practical contexts), the DSP becomes more difficult. Formulating the capacitated version of the problem requires simply replacing each big-M value (M_t) in the DSP formulation with the associated finite capacity level, denoted as b_t, for $t = 1, \ldots, T$. We denote the capacitated demand selection problem by CDSP, and note that, based on the results in the prior section, an equivalent lot sizing with pricing problem with piecewise-linear and concave revenue functions exists. We also observe that, unlike the DSP, the CDSP is an $\mathscr{NP}$-Hard optimization problem, as it contains the capacitated lot sizing problem as a special case. In this section we will briefly discuss modeling issues and problem variants of the CDSP.

Recall that in the DSP formulation it did not matter whether the demand selection (w_{jt}) variables were required to take binary values. That is, even though we formulated the problem such that these variables were free to take any values between zero and one, an optimal solution exists such that each of these variables will be binary at optimality. When finite production capacities exist, however, this is not the case, and whether or not we require these variables to take binary values has a substantial impact on the difficulty of the problem. Moreover, contexts may exist (particularly when capacities are limited) in which not all orders are completely fulfilled, but some are partially fulfilled. We therefore classify capacitated versions of the CDSP according to two important factors. The first of these factors is whether or not partial order fulfillment is permissible. When partial order fulfillment is permissable, the demand selection (w_{jt}) variables are free to take any values between zero and one. If partial order satisfaction is not permitted, then these variables must be binary. The second factor corresponds to whether or not the finite production capacity parameters vary with time. In the equal-capacity case, capacity equals b in every time period, whereas in the general case we retain the subscript on b_t to indicate capacity values that are time-varying.

Although, as noted previously, the CDSP is $\mathscr{NP}$-Hard, the special case with continuous demand selection variables and time-invariant capacities can be solved in polynomial time, based on the polynomial solvability of the equal-capacity lot sizing problem (see [1, 5]). That is, as with the uncapacitated DSP, because the associated problem is polynomially solvable for any given demand vector, and because the demands (orders) in different periods are independent of one another, this special case of the CDSP is polynomially solvable. We next briefly sketch the properties that facilitate its solution in polynomial time.

As with the uncapacitated DSP, an optimal solution exists for the CDSP with equal capacities consisting of a sequence of *regeneration intervals*. A regeneration interval is a subsequence of time periods t through s with $1 \leq t \leq s \leq T$, such that no inventory is held at the end of periods $t-1$ and s, but a positive level of inventory is held at the end of periods $t, t+1, \ldots, s-1$. For the DSP, each arc we defined in the shortest path solution approach corresponded to a regeneration interval, and for the uncapacitated problem, all production in the regeneration interval occurred in the first period (t) of the regeneration interval (defined by the arc $(t, s+1)$). In the equal-capacity version of the lot sizing problem, production within a regeneration interval may occur in any of the periods. However, an optimal solution exists such that, within any regeneration interval, the production level equals either zero or the capacity level b for all periods except at most one (see [1]).

To illustrate why this implies polynomial solvability for the ELSP with constant capacities, note that if we have a regeneration interval $(t, s+1)$, this implies that $I_{t-1}=0$, $I_s=0$, and $I_\tau > 0$ for $\tau = t, t+1, \ldots, s-1$. Because of this, production in periods t through s must satisfy all demands in periods t through s. Letting $D(t,s) = \sum_{\tau=t}^{s} D_\tau$, then if the total production in periods t through s must equal $D(t,s)$, and if production must equal zero or b for all periods except at most one, then we know that the amount produced in the period in which we do not produce at zero or b must equal $D(t,s) \bmod b$, where $x \bmod y = x - y\lfloor x/y \rfloor$. As a result, when solving the subproblem for arc $(t, s+1)$, we only need to consider solutions that (a) produce either zero, b, or $D(t,s) \bmod b$ in each period, and (b) ensure that cumulative production is at least as great as cumulative demand. As a result, for the equal-capacity ELSP, the arc subproblem can be solved in polynomial time, and because the number of arcs is polynomial in T, the overall problem is polynomially solvable.

For the equal-capacity CDSP, these same properties hold for any given demand vector within a regeneration interval, although the number of potential demand vectors is effectively infinite (thus, we do not know the value of $D(t,s)$ *a priori*). A more general characterization of the structural properties of an optimal regeneration interval solution does, however, exist (see [3]). In particular, it is possible to show that an optimal regeneration interval exists that is one of the following types:

1. Production in each period in the regeneration interval equals either 0 or b, and at most one demand is partially satisfied.
2. No demands are partially satisfied, and production may be strictly between 0 and b in at most one period in the regeneration interval.

For any given regeneration interval $(t, s+1)$, we sort all demands in periods t through s in nonincreasing order of $\hat{\rho}_{jt} = \rho_{jt}/d_{jt}$ values. We can show that if $\hat{\rho}_{jt} \geq \hat{\rho}_{kt'}$ for demands j and k in periods t and t' within the regeneration interval, if an optimal solution exists with $w_{jt} < 1$, then an optimal solution exists with $w_{kt'} = 0$; similarly, if $w_{kt'} > 0$ then $w_{jt} = 1$. That is, within any regeneration interval we have a preference ordering of the attractiveness of demands. Then, after sorting in this order, we can evaluate solutions of type 1 above by going down this list and successively assuming that a particular demand is the one such that $0 < w_{jt} < 1$. If a given demand is fractionally satisfied, then all lower indexed demands are fully satisfied, while all higher indexed demands are not satisfied at all. The choice of a particular demand to be satisfied fractionally therefore uniquely determines the number of periods that must be satisfied at full capacity b (assuming $d_{jt} < b$ for all orders), and, as a result, the values of the corresponding w_{jt} variables within the regeneration interval.

For the second type of solution above, given the preference ordering within any regeneration interval, we may consider a total of $\mathcal{O}(n_{\max}T)$ solutions such that demands 1 through l have $w_{jt} = 1$, while the remaining demands have $w_{jt} = 0$. Each such solution implies a value of $D(t,s)$, as well as the number of periods in which production must be at full capacity. We thus have a polynomial number of solutions of types 1 and 2 above that need to be evaluated for each regeneration interval. The time required to determine an optimal subproblem solution is the time required to solve an equal-capacity lot sizing problem (which is $\mathcal{O}(T^3)$) for at most $\mathcal{O}(n_{\max}T)$ demand vectors. This implies that the time required to find an optimal regeneration interval solution is $\mathcal{O}(n_{\max}T^4)$. Because there are $\mathcal{O}(T^2)$ regeneration intervals, the overall time required to solve the equal-capacity CDSP is $\mathcal{O}(n_{\max}T^6)$.

Unfortunately, when we do not permit partial order satisfaction, the resulting problem remains $\mathcal{NP}$-Hard, even when all capacities are equal (see [3]). Methods for strengthening the LP relaxation as well as heuristic solution approaches for this case and problems with time-varying capacities are discussed in [4].

References

1. Florian M, Klein M (1971) Deterministic Production Planning with Concave Costs and Capacity Constraints. Management Science 18:12–20
2. Geunes J, Romeijn H, Taaffe K (2004) Order Selection, Pricing, and Lot Sizing: Models and Solution Algorithms. Technical report 2004-2, Industrial and Systems Engineering Department, University of Florida, Gainesville, FL
3. Geunes J, Romeijn H, Taaffe K (2006) Requirements Planning with Pricing and Order Selection Flexibility. Operations Research 54(2):394–401
4. Taaffe K, Geunes J (2004) Models for Integrated Customer Order Selection and Requirements Planning Decisions under Limited Production Capacity, in Supply Chain and Finance (Pardalos P, Migdalas A, Baourakis G (eds.)) World Scientific, Singapore
5. Van Hoesel C, Wagelmans A (1996) An $\mathcal{O}(T^3)$ Algorithm for the Economic Lot-Sizing Problem with Constant Capacities. Management Science 42(1):142–150
6. Wagelmans A, van Hoesel S, Kolen A (1992) Economic Lot Sizing: An $\mathcal{O}(n \log n)$ Algorithm That Runs in Linear Time in the Wagner-Whitin Case. Operations Research 40(S1):S145–S156

Chapter 6
Dynamic Lot Sizing with Market Selection

Abstract This chapter considers a seemingly innocuous change in the assumptions underlying the Demand Selection Problem (DSP) considered in the previous chapter, which severely complicates the problem analysis. Instead of a sequence of independent demands over a time horizon, in this variant of the problem, demands in successive periods may be related in the sense that if we satisfy a given demand in some period t, we must then satisfy a particular set of demands in other periods. That is, instead of selecting individual demands, we are now faced with the problem of selecting from a set of time-phased vectors of demands. In practical terms, this corresponds to determining whether we will satisfy all or none of a given customer's or market's demands over the time horizon. We refer to the resulting problem as the Market Selection Problem (MSP) and discuss the problem's complexity and potential solution approaches throughout this chapter.

6.1 Market Selection Problem Definition

The Market Selection Problem (MSP) considers a set J of n markets, where the demand from market j in period t equals d_{jt} for all $j \in J$ and $t = 1, \ldots, T$. We can formulate the MSP as a restriction of a special case of the DSP from the previous chapter. That is, starting with the DSP formulation, we assume that $n_t = n$ for all $t = 1, \ldots, T$, where n corresponds to the number of markets. We then impose a new set of constraints that require $w_{jt} = w_{jt+1}$, for all $j \in J$ and $t = 1, \ldots, T-1$. That is, if we choose to satisfy the demand for market j in any period, we must then satisfy that market's demand in every period. This is equivalent to eliminating the time subscript on the w_{jt} variables, and we therefore define w_j as a binary variable equal to one if we satisfy market j demand in all periods, and zero otherwise. We also define R_j as the total revenue available from market j over the T-period horizon, where, using the notation defined in the previous chapter, $R_j = \sum_{t=1}^{T} r_{jt} d_{jt}$. We retain the basic notation and cost assumptions of the DSP except where noted. We can formulate the MSP as follows:

J. Geunes, *Demand Flexibility in Supply Chain Planning*,
SpringerBriefs in Optimization,
DOI 10.1007/978-1-4419-9347-2_6,

$$\text{[MSP]} \quad \text{Maximize} \quad \sum_{j=1}^{n} R_j w_j - \sum_{t=1}^{T} \{S_t y_t + C_t Q_t + H_t I_t\} \tag{6.1}$$

$$\text{Subject to} \quad I_t = Q_t + I_{t-1} - \sum_{j=1}^{n} d_{jt} w_j, \quad t = 1, \ldots, T, \tag{6.2}$$

$$Q_t \le M_t y_t, \quad t = 1, \ldots, T, \tag{6.3}$$

$$I_0 = 0, \quad Q_t, I_t \ge 0, \quad t = 1, \ldots, T, \tag{6.4}$$

$$w_j \in \{0, 1\}, \quad j = 1, \ldots, n, \tag{6.5}$$

$$y_t \in \{0, 1\}, \quad t = 1, \ldots, T. \tag{6.6}$$

Observe that we can set $M_t = \sum_{\tau=t}^{T} \sum_{j=1}^{n} d_{j\tau}$ in (6.3) without loss of optimality. For any given selection of markets, the MSP becomes an ELSP, and is, therefore, easily solved. Similarly, for a given production plan, the problem is easily solved as well. To see this, consider a fixed production plan with υ demands placed in periods $\{1 = t_1 < t_2 < \cdots < t_\upsilon \le T\}$. Because an optimal solution exists satisfying the ZIO property (see Sect. 1.2.3), we know that

$$Q_{t_l} = \sum_{\tau=t_l}^{\tau_{l+1}-1} \sum_{j=1}^{n} d_{j\tau} w_j, \quad l = 1, \ldots, \upsilon, \quad \text{and} \tag{6.7}$$

$$I_t = \sum_{\tau=t+1}^{\tau_{l+1}-1} \sum_{j=1}^{n} d_{j\tau} w_j, \quad t_l \le t < t_l + 1, l = 1, \ldots, \upsilon. \tag{6.8}$$

We can therefore write the net profit associated with a given production plan as

$$\sum_{j=1}^{n} \left(R_j - \sum_{l=1}^{\upsilon} \left\{ C_{t_l} \sum_{\tau=t_l}^{\tau_{l+1}-1} d_{j\tau} + \sum_{\tau=t_l}^{\tau_{l+1}-2} H_\tau \sum_{u=\tau+1}^{\tau_{l+1}-1} d_{ju} \right\} \right) w_j - \sum_{l=1}^{r} S_{t_l}. \tag{6.9}$$

We can maximize (6.9) over all binary vectors w by setting $w_j = 1$ if and only if

$$R_j \ge \sum_{l=1}^{\upsilon} \left\{ C_{t_l} \sum_{\tau=t_l}^{\tau_{l+1}-1} d_{j\tau} + \sum_{\tau=t_l}^{\tau_{l+1}-2} H_\tau \sum_{u=\tau+1}^{\tau_{l+1}-1} d_{ju} \right\}. \tag{6.10}$$

Note that the right-hand side of (6.10) corresponds to the total variable cost incurred when serving market j using the given production plan with υ demands placed in periods $\{1 = t_1 < t_2 < \cdots < t_\upsilon \le T\}$. Thus, this condition simply requires that a market's revenue is at least as great as the associated variable cost incurred in serving the market using the given production plan.

Despite the fact that the problem is easily solved for a given binary vector w or for a given binary vector y, the problem is unfortunately not easily solved when both of these vectors are decision vectors, as we next discuss.

6.1.1 MSP Problem Complexity

The decision version of the MSP was shown to be strongly $\mathscr{N}\mathscr{P}$-Complete in [7]. Although the complete proof of this result is quite involved, here we sketch the proof structure and the main results involved in the proof. We begin by considering a very simple special case of the MSP with an odd number of planning periods T, a fixed order cost of $S_t = 2$ in every period, zero variable production cost in every period (all $C_t = 0$), and a holding cost of one in every period ($H_t = 1$). Consider an instance of the ELSP with these cost parameters and with a demand of $d_t = 1$ in every period. For this ELSP instance, if $v = (T - 1)/2$, then an optimal solution contains $v + 1$ production orders, where v of the orders cover two periods and one of the orders covers one period. The average cost per period of this solution (which also equals the average cost per unit of demand), $AC^*(T)$, equals $3/2 + 1/(2T)$, which clearly exceeds $3/2$.

We next construct a set of markets, such that in every period $t = 1, \ldots, T$, exactly one market has a demand of one unit, and all other markets have a demand of zero units in the period. The revenue associated with market j is defined as

$$R_j = AC^*(T) \sum_{t=1}^{T} d_{jt}. \tag{6.11}$$

Let $J_s \subseteq J$ denote a set of selected markets such that $w_j = 1$ for $j \in J_s$, and let $\Pi(J_s)$ denote the profit associated with the set J_s. Now note that $\Pi(\emptyset) = 0$ and $\Pi(J) = 0$, i.e., selecting either zero markets or all markets produces a profit of zero. Observe also that any selection of markets results in a time-phased demand vector that is binary, and let $\mathscr{D}(J_s)$ denote this vector when the subset J_s of markets is selected. We can thus view any such vector $\mathscr{D}(J_s)$ as a sequence of zeros and ones.

Suppose this vector $\mathscr{D}(J_s)$ contains a subsequence with a zero, followed by an odd number of ones, followed by a zero (for convenience we assume zero demand in dummy periods 0 and $T + 1$). If this is the case, then if k is the total number of ones in the sequence, under our cost assumptions, the total cost equals at least $(3/2)k + (1/2)$. The profit associated with this sequence is no greater than $AC^*(T)k - ((3/2)k + (1/2))$, which is strictly less than $AC^*(T)T - ((3/2)T + (1/2)) = 0$. If, instead, all subsequences of consecutive ones correspond to an even number of periods, then the optimal production plan always uses an order to satisfy two consecutive periods worth of demand, which implies that the average cost per unit produced equals $(3/2)$. In this case, if there are k total ones in the vector $\mathscr{D}(J_s)$, the associated profit equals $(AC^*(T) - (3/2))k > 0$. As a result, a solution with net profit greater than zero exists if and only if the vector $\mathscr{D}(J_s)$ contains only even subsequences of consecutive ones. This implies that the MSP is at least as difficult as the problem of determining whether a selection of markets exists for this special case satisfying this even subsequence property. This even subsequence problem may be posed as follows.

Consider a set J of n binary vectors, each of dimension T, such that for at most one of these vectors, element t equals one. Given a subset $J_s \subseteq J$, let $\mathscr{D}(J_s)$ denote a $(T+2)$-dimensional vector such that elements 1 and $T+2$ equal zero, and elements 2 through $T+1$ consist of the component-wise sum of the vectors in J_s. Note that $\mathscr{D}(J_s)$ must also be a binary vector. We wish to determine whether a subset J_s exists such that every consecutive sequence of ones in $\mathscr{D}(J_s)$ is of even length. The remainder of the proof, detailed in [7], involves a reduction from the classical 3-Satisfiability problem (3SAT; see [1]) to this even subsequence problem. In particular, we consider the market selection recognition problem, which asks whether a selection of markets exists with net profit greater than B for some constant B. Given any instance of 3SAT, [1] creates an equivalent instance of the market selection recognition problem in polynomial time. Thus, if we can answer yes to the market selection recognition problem in polynomial time, this implies that we can answer yes to the question of whether the instance of 3SAT is satisfiable in polynomial time, which only holds if $\mathscr{P} = \mathscr{N}\mathscr{P}$.

6.1.2 MSP Approximability

Because it is not possible to find an optimal solution for an $\mathscr{N}\mathscr{P}$-Hard problem in polynomial time (unless $\mathscr{P} = \mathscr{N}\mathscr{P}$), for such problems, researchers often seek so-called approximation algorithms. An approximation algorithm provides a feasible solution in polynomial time that adheres to a certain provable performance guarantee. For instance, if Π^* denotes the maximum profit of an instance of the MSP, an approximation algorithm A is defined as a $1-\varepsilon$ approximation algorithm for a maximization problem if it is guaranteed to produce a solution with profit Π_A such that $\Pi_A \geq (1-\varepsilon)\Pi^*$. Unfortunately, as shown in [7], a $1-\varepsilon$ approximation algorithm cannot exist for MSP with $0 < \varepsilon < 1$ unless $\mathscr{P} = \mathscr{N}\mathscr{P}$. To see this, suppose that such an algorithm does exist. Recall that in our discussion of the $\mathscr{N}\mathscr{P}$-Completeness proof in the previous section, if no profitable market selection exists (i.e., no selection satisfying the even subsequence property), then $\Pi^* = 0$. This also implies that the corresponding instance of 3SAT is not satisfiable. If such an approximation algorithm exists, then it is guaranteed to find a solution such that $\Pi_A \geq (1-\varepsilon)\Pi^* = 0$, which implies $\Pi_A = 0$. If, on the other hand, $\Pi^* > 0$, then the corresponding instance of 3SAT is indeed satisfiable, and the approximation algorithm finds a solution such that $\Pi_A \geq (1-\varepsilon)\Pi^* > 0$, i.e., $\Pi_A > 0$. But this implies that the approximation algorithm can verify, in polynomial time, whether the instance of 3SAT is satisfiable, which would imply that 3SAT is polynomially solvable (we thus have a contradiction, unless $\mathscr{P} = \mathscr{N}\mathscr{P}$).

Suppose that we instead consider an alternate approach to formulating the MSP, as discussed in [2] and [4]. In particular, suppose that we view R_j as the opportunity cost for not satisfying market j demand, and let $\bar{w}_j = 1 - w_j$. Thus, if $\bar{w}_j = 1$, we forgo the revenue associated with market j demand and we need not satisfy market j demand. We also reformulate the constraint set using a facility location type of reformulation of the constraint set of the MSP. That is, we let $x_{jt\tau}$ denote the *percentage* of market j demand in period τ satisfied using production in period t, and we define $c^j_{t\tau} = (C_t + \sum_{s=t}^{\tau-1} H_s)d_{jt}$ as the cost to satisfy market j demand in period τ using production in period t. Then we can formulate the LP relaxation of the MSP as follows:

$$\text{[MSP-alt]} \quad \text{Minimize} \quad \sum_{j=1}^{n} R_j \bar{w}_j + \sum_{t=1}^{T} \left\{ S_t y_t + \sum_{j=1}^{n} \sum_{\tau=t}^{T} c^j_{t\tau} x_{jt\tau} \right\} \tag{6.12}$$

$$\text{Subject to} \quad \sum_{t=1}^{\tau} x_{jt\tau} + \bar{w}_j = 1, \quad \tau = 1, \ldots, T, \; j \in J, \tag{6.13}$$

$$x_{jt\tau} \leq y_t, \quad t = 1, \ldots, T, \; \tau = t, \ldots, T, \; j \in J, \tag{6.14}$$

$$x_{jt\tau}, y_t \geq 0, \quad t = 1, \ldots, T, \; \tau = t, \ldots, T, \; j \in J. \tag{6.15}$$

The first term of the objective function (6.12) corresponds to the market rejection costs, while the remaining terms in brackets are the associated lot sizing costs. Consider an optimal solution $(\hat{w}, \hat{x}, \hat{y})$ for the MSP-alt formulation. Now suppose that for any market such that $\hat{w}_j \geq 1/2$, we set $\bar{w}_j = 1$, and for all other markets, we set $\bar{w} = 0$. Let $J_{1/2} = \{j : \hat{w}_j \geq 1/2\}$; this implies that $\bar{w}_j = 1$ for $j \in J_{1/2}$ and $\bar{w}_j = 0$ for $j \in J \backslash J_{1/2}$. We refer to the resulting $\bar{w}$ as the rounded LP relaxation market selection vector, and we denote this vector as $\bar{w}^r$.

The *rejection cost* associated with the rounded market selection vector $\bar{w}^r$ equals $\sum_{j \in J_{1/2}} R_j$. The rejection cost associated with the LP relaxation solution equals $\sum_{j \in J} R_j \hat{w}_j$, and we have $2 \times \sum_{j \in J_{1/2}} R_j \hat{w}_j \geq \sum_{j \in J_{1/2}} R_j$ and $2 \times \sum_{j \in J \backslash J_{1/2}} R_j \hat{w}_j \geq 0$, which implies that $\sum_{j \in J_{1/2}} R_j \leq 2 \times \sum_{j \in J} R_j \hat{w}_j$. That is, the rejection cost of the rounded market selection vector is at most twice the rejection cost associated with the LP relaxation solution.

We next demonstrate that by solving the ELSP that results when fixing $\bar{w} = \bar{w}^r$, we arrive at a feasible solution for the MSP with total cost no greater than twice the cost of the LP relaxation of MSP-alt. Because the LP relaxation of MSP-alt provides a lower bound on the optimal solution of the MSP, this implies we can obtain a feasible solution for the MSP with cost at most twice that of the optimal solution.

Let ELSP(1/2) denote the instance of the ELSP that results when the vector $\bar{w}$ is fixed at $\bar{w}^r$. When we solve MSP-alt with $\bar{w} = \bar{w}^r$, we know that an optimal solution exists for the resulting LP relaxation with lot sizing costs equal to the optimal solution value of ELSP(1/2) (because no integrality gap exists for this formulation of the ELSP). Thus, if we can find a solution to the restricted version of MSP-alt when $\bar{w} = \bar{w}^r$ with lot sizing costs at most twice those associated with the LP relaxation solution $(\hat{w}, \hat{x}, \hat{y})$, then we know that ELSP(1/2) has an optimal solution value no more than twice the lot sizing costs associated with $(\hat{w}, \hat{x}, \hat{y})$. This implies that the total rejection plus lot sizing costs of the feasible solution we have constructed are no more than twice the LP relaxation solution.

For $j \in J_{1/2}$, let $x_{jt\tau} = 0$ for all $t = 1, \ldots, T$ and $\tau = t, \ldots, T$. For $j \in J \backslash J_{1/2}$, let $\tilde{x}_{jt\tau} = \frac{\hat{x}_{jt\tau}}{1-\hat{w}_j}$, and for every $t = 1, \ldots, T$, let $\tilde{y}_t = \max\{\min_{j \in J \backslash J_{1/2}}\{\frac{\hat{y}_t}{1-\hat{w}_j}\}, 1\}$. Note that this solution is feasible for MSP-alt and that for $j \in J \backslash J_{1/2}$, we have $\tilde{x}_{jt\tau} \leq 2\hat{x}_{jt\tau}$ for all $t = 1, \ldots, T$ and $\tau = t, \ldots, T$, and for $t = 1, \ldots, T$, $\tilde{y}_t \leq \hat{y}_t$. Letting $\bar{x}, \bar{y}$ denote an optimal solution for ELSP(1/2), we then have

$$\sum_{t=1}^{T}\left\{S_t \bar{y}_t + \sum_{j=1}^{n}\sum_{\tau=t}^{T} c_{t\tau}^j \bar{x}_{jt\tau}\right\} \leq \sum_{t=1}^{T}\left\{S_t \tilde{y}_t + \sum_{j=1}^{n}\sum_{\tau=t}^{T} c_{t\tau}^j \tilde{x}_{jt\tau}\right\}$$
$$\leq 2 \times \sum_{t=1}^{T}\left\{S_t \hat{y}_t + \sum_{j=1}^{n}\sum_{\tau=t}^{T} c_{t\tau}^j \hat{x}_{jt\tau}\right\}.$$

We have thus shown that we can construct a feasible solution $(\bar{w}^r, \bar{x}, \bar{y})$ with rejection costs at most twice those of the LP relaxation solution, and lot sizing costs at most twice those of the LP relaxation solution, which implies an approximation algorithm for the MSP with a performance guarantee of no more than twice the optimal solution (with respect to the objective function (6.12)). Note, however, that our prior result for the MSP formulation still holds, i.e., a $1-\varepsilon$ approximation algorithm cannot exist for MSP with $0 < \varepsilon < 1$ unless $\mathscr{P} = \mathscr{NP}$.

To see this, for any selection of markets $J_s \subseteq J$, let $\Gamma(J_s)$ equal the minimum cost associated with this subset according to (6.12), and let $\Pi(J_s)$ equal the maximum net profit according to (6.1). We then always have $\Gamma(J_s) + \Pi(J_s) = \sum_{j=1}^n R_j$. That is, if we substitute $1 - w_j = \bar{w}_j$ for all $j \in J$ in (6.12), we obtain the constant term $\sum_{j=1}^n R_j$ within the objective. Now observe that if J_s^* is an optimal selection of markets and J_s^A is a selection of markets found by a $1+\varepsilon$ approximation algorithm for MSP-alt, then $\Gamma(J_s^A)/\Gamma(J_s^*) \leq 1+\varepsilon$, which is equivalent to

$$\frac{\sum_{j=1}^n R_j - \Pi(J_s^A)}{\sum_{j=1}^n R_j - \Pi(J_s^*)} \leq 1+\varepsilon, \tag{6.16}$$

which is equivalent to

$$\frac{\Pi(J_s^A)}{\Pi(J_s^*)} \geq 1 - \varepsilon\left(\frac{\sum_{j=1}^n R_j}{\Pi(J_s^*)} - 1\right). \tag{6.17}$$

Because $\sum_{j=1}^n R_j$ can be arbitrarily large, a cost performance guarantee for MSP-alt does not imply a corresponding performance guarantee for the MSP.

We next note that our choice of $1/2$ as the threshold for rounding was somewhat arbitrary and, based on previous approaches that demonstrate the success of randomized algorithms (see [5, 6]), we can devise a similar randomized method. Instead of using $1/2$ as our rounding threshold, suppose we use β for some value chosen randomly from the interval $(0, \delta]$. Thus, if $\hat{w}_j \geq \beta$, we reject market j, and we select market j otherwise. Because market j will be rejected if $\hat{w}_j \geq \beta$, the probability we reject market j is the probability that β is less than or equal to $\hat{w}_j$ for β selected uniformly from the interval $(0, \delta]$. Therefore, if $\hat{w}_j \leq \delta$, the probability that market j is rejected is no higher than $\hat{w}_j/\delta$. If $\hat{w}_j > \delta$, then we reject market j with probability 1. Noting that when $\hat{w}_j > \delta$, the fraction $\hat{w}_j/\delta > 1$, this implies that the probability of rejecting market j is always bound from above by $\hat{w}_j/\delta$. We can therefore bound the total rejection cost from above using the term $\sum_{j \in J} R_j \hat{w}_j/\delta$. Using a similar approach as we did for the case in which the rounding threshold was $1/2$, we can show that this rounding approach using a threshold of β produces a solution with lot sizing costs that are at most $1/(1-\beta)$ times the lot sizing costs from the LP relaxation. To get the expected lot sizing costs from a randomly chosen value of β from the interval $(0, \delta]$, we take the integral $(1/\delta)\int_0^\delta [1/(1-\beta)]\,d\beta$, which equals $\ln(1/(1-\delta))/\delta$. As a result, the *expected* lot sizing costs we obtain when solving ELSP(β) are no higher than $\ln(1/(1-\delta))/\delta$ times the lot sizing costs from the LP relaxation. We are thus able to create a solution with rejection costs at most $1/\delta$ times the rejection costs of the LP relaxation, and with expected lot sizing costs of at most $\ln(1/(1-\delta))/\delta$ times the lot sizing costs from the LP relaxation. Observe that when $\delta = 1 - e^{-1}$ we have $1/\delta = \ln(1/(1-\delta))/\delta < 1.582$.

We can next *derandomize* the algorithm using the following approach. Because the randomly chosen value of β results in some precise set of selected markets, this implies there are n intervals of β values such that any value chosen from an interval results in the same solution in terms of the markets selected and rejected. These intervals are completely specified by the $\hat{w}_j$ values. Thus, we can select any value from each one of these intervals and apply the rounding algorithm associated with each of these values. The minimum cost solution among these will have cost no greater than the expected cost, which implies that we have an approximation algorithm with a performance guarantee of 1.582.

6.1.3 Polynomially Solvable Special Cases

Despite the negative complexity results we have discussed, a number of special cases of the MSP turn out to be solvable in polynomial time, several of which are not unlikely to arise in practice. This section briefly summarizes a class of polynomially solvable special cases of the MSP.

When contrasting the MSP with the DSP, it is apparent that the difference in complexity lies in the interdependence of each market's demands across time periods, because satisfying a market's demand in one period implies the need to satisfy its demand in all periods. This eliminates the possibility of applying a dynamic programming (shortest path) approach that decomposes the problem into a sequence of smaller problems with shorter time horizons. As a result, even for cases with special structure, we must identify alternative solution approaches.

We first consider the simple case in which market demands are time-invariant. That is, for each market j, we assume $d_{jt} = d_j$ for $t = 1, \ldots, T$. In this case, we can show that a similar sorting algorithm to the one used for the EOQMC in Sect. 3.1.1 can be used to solve the problem. This result is stated in the following lemma. Before presenting the lemma, we define $D^j = Td_j$ as the total market j demand over the planning horizon.

Lemma 6.1 *If market demands are time-invariant and markets are indexed in non-increasing order of unit revenue, R_j/D_j, values, then if an optimal solution exists that selects market k, an optimal solution exists that selects market $k-1$.*

Proof Suppose we have an optimal solution in which market k is selected and market $k-1$ is not. Let J' denote the set of selected markets in this solution, and let $\mathscr{R}$ equal the total revenue obtained in this solution except for market k, i.e., $\mathscr{R} = \sum_{j \in J' \setminus \{k\}} R_j$. Define $\Gamma^*(J' \setminus \{k\})$ as the minimum total lot sizing costs associated with the solution we have defined, less the total variable production and holding costs associated with market k, which we denote as p_k and h_k, respectively. The net profit associated with the solution we have defined equals $\mathscr{R} + R_k - \Gamma^*(J' \setminus \{k\}) - p_k - h_k$. Because market k is selected in an optimal solution, we must have $\mathscr{R} + R_k - \Gamma^*(J' \setminus \{k\}) - p_k - h_k \geq \mathscr{R} - \Gamma^*(J' \setminus \{k\})$, or $R_k \geq p_k + h_k$. Next consider adding market $k-1$ to the optimal solution we have defined, and let p_{k-1} and h_{k-1} equal the production and holding costs associated with market $k-1$ when using the same set of order periods as our optimal solution. Because production and holding costs are linear and market demands are time-invariant, we must have $\frac{p_k + h_k}{D_k} = \frac{p_{k-1} + h_{k-1}}{D_{k-1}}$. But our initial sort order implies that $\frac{R_{k-1}}{D_{k-1}} \geq \frac{R_k}{D_k} \geq \frac{p_k + h_k}{D_k} = \frac{p_{k-1} + h_{k-1}}{D_{k-1}}$, which implies $R_{k-1} \geq p_{k-1} + h_{k-1}$. As a result, if we add market $k-1$ to the set of selected markets, then the resulting solution is at least as good as our initial solution, i.e., either we have a contradiction or an alternative optimal solution. □

Lemma 6.1 implies that we can first sort markets in nonincreasing order of R_j/D_j, then solve a sequence of n lot sizing problems, where problem k selects markets $1, \ldots, k$, in order to find an optimal solution for the problem with time-invariant market demands.

Corollary 6.1 *When market j demand in period t equals a base market demand level d_j multiplied by a seasonal factor σ_t, i.e., $d_{jt} = \sigma_t d_j$, then Lemma 6.1 remains valid, with $D_j = \sum_{t=1}^{T} \sigma_t d_j$.*

Note that because these cases use a ratio that divides R_j by Td_j in the first case, and by $d_j(\sum_{t=1}^{T} \sigma_t)$ in the seasonal case, both of these cases are equivalent to sorting in nonincreasing order of R_j/d_j. Because we need to first sort n markets and then solve n instances of the ELSP, the complexity associated with this problem is $\mathcal{O}(n(\log n + T \log T))$.

In [7], a number of additional polynomially solvable special cases are discussed, which we briefly describe in the following list.

1. Market-specific pricing, with market j demand in period t taking the form $d_{jt} = \alpha_t - \beta_t p_j$, where α_t and β_t are constants and p_j is the market j price.
2. Infinite holding costs, in which case the problem becomes equivalent to a class of shared fixed cost and selection problems discussed in [3].
3. Instances in which each market has positive demands spanning at most k consecutive periods (the DSP is the special case in which $k = 1$).
4. Instances containing a staircase demand matrix, where each period has demand from at most one market, and each market's positive demands span a continuous set of periods.

It is interesting to observe that for the MSP with time-invariant (and seasonal) demands, the sorting rule is based strictly on the markets' unit revenues, which is precisely the sorting rule used for the EOQMC in Chap. 3.

6.1.4 Heuristic Solution Methods

Clearly the LP rounding approach discussed in Sect. 6.1.2 provides a heuristic solution approach with a worst-case performance ratio of 1.582. An additional simple heuristic approach is described in [7], which we next summarize. This heuristic applies an iterative approach that takes advantage of the fact that for either a given set of markets or for a given set of production periods, the remaining problem is easily and quickly solved. To define this heuristic approach, let $y^*(w)$ denote an optimal value of the vector y for a given market selection vector w. Similarly, let $w^*(y)$ denote an optimal market selection vector w for a given vector y. Starting with a specific order plan defined by a vector y^0, we use condition (6.10) to determine an optimal selection of markets $w^0 = w^*(y^0)$. Next, given $w = w^0$, we solve an instance of the ELSP to obtain $y^1 = y^*(w^0)$. More generally, given a vector y^i at iteration i, we determine $w^i = w^*(y^i)$ and then obtain $y^{i+1} = y^*(w^i)$, continuing until either $y^{i+1} = y^i$ or $w^{i+1} = w^i$. The net profit is guaranteed to improve at each iteration (except the last) and the algorithm is finite because the number of possible order plans is finite. While this iterative solution approach is quite simple and fast, the quality of the final (locally optimal) solution obtained is very sensitive to the initial solution y^0. It is thus important to run the algorithm using numerous initial solutions as a starting point, as described in [7]. As shown in [7], this iterative algorithm is extremely effective in terms of average performance based on a large number of computational test instances.

References

1. Garey M, Johnson D (1979) Computers and Intractability. W.H. Freeman and Company, New York
2. Geunes J, Levi R, Romeijn H, Shmoys D (2011) Approximation Algorithms for Supply Chain Planning and Logistics Problems with Market Choice. Mathematical Programming, Series A 130(1):85–106
3. Hochbaum D (2004) Selection, Provisioning, Shared Fixed Costs, Maximum Closure, and Implications on Algorithmic Methods Today. Management Science 50(6):709–723
4. Levi R, Geunes J, Romeijn H, Shmoys D (2005) Inventory and Facility Location Models with Market Selection. Lecture Notes in Computer Science 3509(2005):237–259
5. Motwani R, Naor J, Raghavan P (1996) Randomized Approximation Algorithms in Combinatorial Optimization, in Approximation Algorithms for NP-hard Problems (Hochbaum D (ed.)) Thomson, Boston, 447–481
6. Raghavan P, Thompson C (1987) Randomized Rounding: A Technique for Provably Good Algorithms and Algorithmic Proofs. Combinatorica 7:365–374
7. Van den Heuvel W, Kundakcioglu E, Geunes J, Romeijn H, Sharkey T, Wagelmans A (2011) Integrated Market Selection and Production Planning: Complexity and Solution Approaches. Mathematical Programming, Series A (forthcoming)

Part III
Supply Chain Network Planning with Demand Flexibility

Chapter 7
Assignment and Location Problems in Supply Chains

Abstract This chapter delves into problems that require assigning customer demands to supply sources. When no fixed cost exists for using a supply source and supply sources are capacitated, we have a generalized assignment problem (GAP). When fixed costs are incurred for using a supply source, then the problem is a classical facility location problem (FLP). We consider both capacitated and uncapacitated facility location problems, as well as the implications of requiring single sourcing constraints that do not allow splitting a demand between supply sources. Within these problem classes, we analyze two different forms of demand flexibility. The first type of flexibility corresponds to what we saw in the last two chapters, i.e., each demand must be either fully accepted or rejected. The second type of flexibility requires satisfying each demand, but the quantity (e.g., size, number of units) at which a demand is satisfied must fall between prespecified lower and upper limits, while revenue is proportional to the level at which the demand is satisfied. This chapter defines several such assignment and location models containing these dimensions of demand flexibility.

7.1 Demand Selection Problems

The models we present in this section follow naturally from the work discussed in Chaps. 5 and 6, based on the classical GAP and FLP models defined in Sects. 1.2.5 and 1.2.6, respectively.

7.1.1 The GAP with Demand Selection

Recall that our standard definition of the GAP in Sect. 1.2.5 sought to minimize the cost associated with the assignment of a set of demands J to a set of resources I, where the cost of assigning demand $j \in J$ to resource $i \in I$ equals c_{ij}. In the demand selection version of the GAP, instead of minimizing cost, we maximize profit, where π_{ij} denotes the profit associated with assigning demand j to resource i. Instead of requiring the assignment of each demand to a resource, as in constraint set (1.21) of the GAP formulation, we allow assigning each demand j to *at most* one resource i

J. Geunes, *Demand Flexibility in Supply Chain Planning*,
SpringerBriefs in Optimization,
DOI 10.1007/978-1-4419-9347-2_7,

(in this sense, this demand selection version of the GAP departs from the standard definition of a pure assignment problem), which results in the following formulation of the GAP with demand selection:

$$\text{[GAPDS]} \quad \text{Maximize} \quad \sum_{i=1}^{m} \sum_{j=1}^{n} \pi_{ij} x_{ij} \tag{7.1}$$

$$\text{Subject to} \quad \sum_{j=1}^{n} D_{ij} x_{ij} \leq b_i, \quad i = 1, \ldots, m, \tag{7.2}$$

$$\sum_{i=1}^{m} x_{ij} \leq 1, \qquad j = 1, \ldots, n, \tag{7.3}$$

$$x_{ij} \in \{0, 1\}, \qquad i = 1, \ldots, m, \; j = 1, \ldots, n. \tag{7.4}$$

The GAPDS is a slight generalization of the standard multiple knapsack problem (see [2]), where the above formulation permits both the revenue π_{ij} and capacity consumption D_{ij} values to depend on the resource to which a demand is assigned (while the standard multiple knapsack problem would use parameters π_j and D_j, independent of the resource). This implies that the GAPDS is an $\mathcal{NP}$-Hard optimization problem, although numerous effective solution methods have been developed for the multiple knapsack problem.

7.1.2 The FLP with Demand Selection

This section describes a generalization of the FLP that is similar to the way in which the GAP was generalized in the previous section. That is, we let π_{ij} denote the net revenue if demand j is assigned to facility i, and we no longer require satisfying every demand, as was required in the standard definition of the FLP in Sect. 1.2.6:

$$\text{[FLPDS]} \quad \text{Maximize} \quad \sum_{i=1}^{m} \sum_{j=1}^{n} \pi_{ij} x_{ij} - \sum_{i=1}^{m} S_i y_i \tag{7.5}$$

$$\text{Subject to} \quad \sum_{j=1}^{n} D_j x_{ij} \leq b_i y_i, \quad i = 1, \ldots, m, \tag{7.6}$$

$$\sum_{i=1}^{m} x_{ij} \leq 1, \qquad j = 1, \ldots, n, \tag{7.7}$$

$$x_{ij} \in \Omega, \tag{7.8}$$

$$y_i \in \{0, 1\}, \qquad i = 1, \ldots, m. \tag{7.9}$$

Recall that Ω is either $\{0, 1\}^{m \times n}$ (when single sourcing is required) or $[0, 1]^{m \times n}$, when splitting any demand across multiple sources is permitted (also recall that this distinction is unimportant when resources are uncapacitated, as an optimal single-sourcing solution exists when the vector x is continuous). The FLPDS is $\mathcal{NP}$-Hard, even when resources are uncapacitated, as shown in [3]. The FLPDS also generalizes the DSP in the same way that the FLP generalizes the ELSP. A heuristic dual ascent algorithm for the uncapacitated version of the FLPDS is provided in [4].

7.2 Problems with Demand Specification Flexibility

In this section we define models that permit an even more general level of demand flexibility than exists in the models we have defined so far throughout this book. We indirectly alluded to this different type of flexibility in Sect. 5.5, when we permitted partial demand satisfaction, but until now we have not dealt with it directly. We note that the FLPDS when $\Omega = [0, 1]^{m \times n}$ also allows partial demand satisfaction (although in the uncapacitated version of this problem it is possible to show that an optimal solution exists in which such partial demand satisfaction does not occur), and is thus a special case of the kind of demand flexibility that we next define, which we refer to as *demand specification flexibility*.

Under demand specification flexibility, for a given demand indexed by j, we must satisfy demand at a level between some prespecified lower and upper bounds, l_j and u_j, if we must meet the demand. In describing the majority of models we have dealt with so far, we have used a single parameter (or vector) D_j to define the quantity of demand that must be met in order to satisfy demand j. The parameter (or vector) D_j *specifies* the demand level associated with demand j, while demand specification flexibility allows the demand level associated with demand j to vary between some lower and upper limits. In the selection models we have discussed so far, the quantity of demand j that is satisfied must equal either 0 or D_j. Thus, if w_j denotes a variable that must equal one if we select demand j and zero otherwise, then $D_j w_j$ corresponds to the quantity of demand j that is satisfied.

Under demand specification flexibility, on the other hand, we define a new variable v_j associated with demand j, which specifies the level at which demand j will be satisfied. Given the lower and upper bounds l_j and u_j, we will require $l_j \le v_j \le u_j$, assuming that demand j must be satisfied. If we do not require satisfying demand j, then we can utilize the binary variable w_j along with the adjusted constraint $l_j w_j \le v_j \le u_j w_j$ to allow for not satisfying demand j at all. The case in which $l_j = u_j = D_j$ corresponds to the majority of the selection-type models we have dealt with thus far, where we can then write $v_j = D_j w_j$. For the limited number of models we have discussed that permit partial demand satisfaction, we have $l_j = 0$, $u_j = D_j$, and $v_j = D_j w_j$, where the variable w_j is now allowed to take any value on the interval $[0, 1]$.

Demand specification flexibility arises in a number of practical settings, especially when interpreted as partial demand satisfaction, which imposes a lower bound of zero on v_j. Cases with non-zero lower bounds also arise in practice, particularly in continuous manufacturing processes of raw materials (e.g., metal or plastic strips, plates, and tubes, or wood products used in further construction, manufacturing, distribution, and finishing operations). In such settings, because of the inherent value of the material or product in question, customers often pay a price that is proportional to the quantity (e.g., weight, length, surface area) delivered, or a price per unit that is increasing at a decreasing rate (i.e., a concave function) in the quantity delivered. As a result, in addition to defining the variable v_j, we also define a revenue function $R_j(v_j)$ that characterizes the total revenue obtained by the supplier (from the customer) when demand j is delivered at the level v_j.

7.2.1 The GAP with Demand Specification Flexibility

Our definition of the GAP with flexible demands (GAPFD) allows lower and upper bounds on each demand level j, which are resource dependent, i.e., if demand j is allocated to resource i, then the delivered quantity for demand j, v_{ij}, must fall between lower and upper bounds l_{ij} and u_{ij}. Let x_{ij} equal one if demand j is allocated to resource i, and zero otherwise. We assume a revenue function of the form $R_{ij}(v_{ij}) = \pi_{ij}x_{ij} + r_{ij}v_{ij}$ (note that our constraints will require that $v_{ij} > 0$ implies $x_{ij} = 1$, and the revenue function can therefore be expressed as a function of v_{ij} alone). Using this function, the parameter π_{ij} corresponds to the fixed profit obtained if resource i is used to satisfy demand j, while r_{ij} is the corresponding revenue per unit of demand. We also assume that a fixed amount of capacity a_{ij} is consumed when assigning demand j to resource i (due to, e.g., resource setup time), in addition to the amount of capacity consumed due to the quantity delivered, v_{ij}. We formulate the GAPFD as follows:

$$\text{[GAPFD]} \quad \text{Maximize} \quad \sum_{i=1}^{m}\sum_{j=1}^{n}\{\pi_{ij}x_{ij} + r_{ij}v_{ij}\} \tag{7.10}$$

$$\text{Subject to} \quad \sum_{j=1}^{n}\{a_{ij}x_{ij} + v_{ij}\} \le b_i, \quad i = 1, \ldots, m, \tag{7.11}$$

$$\sum_{i=1}^{m} x_{ij} = 1, \quad j = 1, \ldots, n, \tag{7.12}$$

$$l_{ij}x_{ij} \le v_{ij} \le u_{ij}x_{ij}, \quad i = 1, \ldots, m, \; j = 1, \ldots, n, \tag{7.13}$$

$$x_{ij} \in \{0, 1\}, \quad i = 1, \ldots, m, \; j = 1, \ldots, n. \tag{7.14}$$

The problem when the assignment constraints (7.12) are cast as less-than-or-equal-to inequalities (as in the GAPDS) serves as a relaxation of the GAPFD, although we focus on the problem as formulated above, with equality constraints.

We will next briefly describe an effective heuristic solution approach for the GAPFD detailed in [6]. To motivate this heuristic, note that the following linear program is equivalent to the LP relaxation of the GAPFD (see [6]):

$$[\text{LP}'] \quad \text{Maximize} \quad \sum_{i=1}^{m}\sum_{j=1}^{n}\left\{(\pi_{ij}+r_{ij}u_{ij})s_{ij}+(\pi_{ij}+r_{ij}l_{ij})t_{ij}\right\} \tag{7.15}$$

$$\text{Subject to} \quad \sum_{j=1}^{n}\left\{(a_{ij}+u_{ij})s_{ij}+(a_{ij}+l_{ij})t_{ij}\right\}\le b_i, \quad i=1,\dots,m, \tag{7.16}$$

$$\sum_{i=1}^{m} s_{ij}+t_{ij}=1, \quad j=1,\dots,n, \tag{7.17}$$

$$s_{ij},t_{ij}\ge 0, \quad i=1,\dots,m, \; j=1,\dots,n. \tag{7.18}$$

Let λ_i, $i \in I$, and μ_j, $j \in J$ denote dual variables associated with constraints (7.16) and (7.17). Then the dual of LP′, which we denote as D′, is formulated as follows:

$$[\text{D}'] \quad \text{Minimize} \quad \sum_{i=1}^{m}\lambda_i bi+\sum_{j=1}^{n}\mu_j \tag{7.19}$$

$$\text{Subject to} \quad \mu_j \ge \pi_{ij}-\lambda_i a_{ij}+(r_{ij}-\lambda_i)l_{ij}, \quad i=1,\dots,m,\; j=1,\dots,n, \tag{7.20}$$

$$\mu_j \ge \pi_{ij}-\lambda_i a_{ij}+(r_{ij}-\lambda_i)u_{ij}, \quad i=1,\dots,m,\; j=1,\dots,n, \tag{7.21}$$

$$\lambda_i \ge 0, \quad i=1,\dots,m, \tag{7.22}$$

$$\mu_j \text{ unrestricted}, \quad j=1,\dots,n. \tag{7.23}$$

Note that constraints (7.20), (7.21), and (7.20) imply

$$\mu_j=\pi_{ij}-a_{ij}\lambda_i+\max\left\{(r_{ij}-\lambda_i)l_{ij};(r_{ij}-\lambda_i)u_{ij}\right\}, \quad i=1,\dots,m,\; j=1,\dots,n, \tag{7.24}$$

where the inner max equals $(r_{ij}-\lambda_i)u_{ij}$ when $\lambda_i \le r_{ij}$ and equals $(r_{ij}-\lambda_i)l_{ij}$ otherwise. If we define

$$f_\lambda(i,j)=\pi_{ij}-a_{ij}\lambda_i+\begin{cases}(r_{ij}-\lambda_i)u_{ij}, & \lambda_i \le r_{ij},\\ (r_{ij}-\lambda_i)l_{ij}, & \lambda_i > r_{ij},\end{cases} \tag{7.25}$$

then we can write $\mu_j = \max_{i=1,\ldots,m} f_\lambda(i, j)$ for $j = 1, \ldots, n$. The dual problem may therefore be written as

$$\min_{\lambda \geq 0} \sum_{j=1}^{n} \max_{i=1,\ldots,m} f_\lambda(i, j) + \sum_{i=1}^{m} \lambda_i b_i. \tag{7.26}$$

We can view Eq. (7.25), and thus $f_\lambda(i, j)$, as the *pseudo-profit* associated with the assignment of demand j to resource i for a given value of the dual multiplier λ_i. The term $\pi_{ij} - a_{ij}$ is a fixed net pseudo-profit term, while the last term in (7.25) may be interpreted as the additional net profit per unit of capacity consumption, multiplied by the amount of capacity consumption. Thus, if $\lambda_i \leq r_{ij}$, then the assignment of demand j to resource i is profitable, and if we make this assignment, we should do so at the upper bound, u_{ij}, if possible. Similarly, if $\lambda_i > r_{ij}$, then assigning demand j to resource i provides a negative contribution to profit per unit of capacity consumption, and if we make this assignment, we should do so at a value no greater than the lower bound l_{ij}.

The heuristic solution method provided in [6] uses the pseudo-profit functions as the basis for a greedy heuristic solution approach (unless specifically noted, we assume that the value of the vector λ used in the heuristic is determined by an optimal dual solution to D′). To describe this heuristic, we first define i_j as the most profitable resource for demand j, i.e., $i_j = \arg\max_{i \in I} f_\lambda(i, j)$. The greedy approach considers not only the most profitable resource for each demand j, but also the difference between this value and the second most profitable resource for the demand. That is, we define

$$\hat{f}_j = f_\lambda(i_j, j) - \max_{i' \in I \setminus \{i_j\}} f_\lambda(i', j), \tag{7.27}$$

in order to capture the desirability of assigning demand j to resource i_j. This measure is used for capacitated problems to get an idea of how critical it is to assign a demand to its most profitable resource. If $\hat{f}_j$ is small, then there is not much loss in assigning demand j to its second most profitable resource, and if $\hat{f}_j$ is large, then a significant penalty exists if we cannot make this assignment. In the greedy phase of the heuristic, the algorithm considers demands in decreasing order of $\hat{f}_j$ values and makes the assignments that are feasible to the most attractive resource for each demand. If an assignment of demand j to resource i is feasible at a value at least equal to the lower bound l_{ij} and $r_{ij} > \lambda_i$, then the assignment is made at the largest value of v_{ij} possible (the minimum between the remaining available capacity of resource i and u_{ij}); if $r_{ij} \leq \lambda_i$, then the assignment is made at $v_{ij} = l_{ij}$, if possible. If the greedy phase of the algorithm terminates with unassigned demands, then an improvement phase is implemented in an attempt to find a feasible solution that assigns all demands to resources.

Some important structural properties of the optimal solution of LP′ and the heuristic approach we have outlined are provided in [6], which permit deriving key performance characteristics and guarantees for this heuristic approach. We next summarize these key properties and the results they imply. The first key result states that if the solution to LP′ is unique and we use the corresponding dual solution vector λ^* in the heuristic approach, then for all demands in the solution to LP′ that are not split between two or more resources, the greedy phase of the heuristic assigns these demands to the exact same resources as the solution to LP′. Moreover, for such demands that are fulfilled at their upper or lower limits in LP′, the heuristic assigns these demands at the same level of resource consumption as in the solution to LP′. The next key property shows that the total number of demands that are either split between multiple resources or are fulfilled at a level strictly between their upper and lower limits in the solution to LP′ is bounded from above by m, the number of resources. These results permit obtaining asymptotic performance results for the heuristic for problems such that the number of resources is held fixed and the number of demands and resource capacities increase to infinity (where resource capacities increase linearly in the number of demands in a way that ensures feasibility with probability one). In particular, under mild assumptions on the probability distributions that characterize the problem parameters, and under specific assumptions on the growth of resource capacities as the number of demands increases, [6] shows that the heuristic approach we have outlined is asymptotically feasible and optimal.

7.2.2 The FLP with Demand Specification Flexibility

The facility location problem with demand specification flexibility, FLPFD, generalizes the GAPFD in the same way the FLP generalizes the GAP, by including fixed costs for the use of resources. Thus, letting y_i denote a binary variable equal to one if resource i satisfies any demand, and zero otherwise, we formulate the FLPFD as follows:

$$\text{[FLPFD]} \quad \text{Maximize} \quad \sum_{i=1}^{m}\sum_{j=1}^{n}\{\pi_{ij}x_{ij} + r_{ij}v_{ij}\} - \sum_{i=1}^{m} S_i y_i \tag{7.28}$$

$$\text{Subject to} \quad \sum_{j=1}^{n}\{a_{ij}x_{ij} + v_{ij}\} \le b_i y_i, \quad i = 1,\ldots,m, \tag{7.29}$$

$$\sum_{i=1}^{m} x_{ij} = 1, \quad j = 1,\ldots,n, \tag{7.30}$$

$$l_{ij}x_{ij} \le v_{ij} \le u_{ij}x_{ij}, \quad i = 1,\ldots,m, \; j = 1,\ldots,n, \tag{7.31}$$

$$x \in \Omega, \tag{7.32}$$

$$y_i \in \{0, 1\}, \quad i = 1,\ldots,m. \tag{7.33}$$

Recall that when $\Omega = \{0, 1\}^{m\times n}$, we have the single-sourcing version of the problem, and when $\Omega = [0, 1]^{m\times n}$, this requirement is relaxed and demands may be split among facilities. The FLPFD problem is formulated using a more general revenue function $r_{ij}(v_{ij})$ in [5] instead of the linear revenue function $r_{ij}v_{ij}$ we use above. In [5], an exact branch-and-price solution method is provided for the FLPFD, while a very-large-scale-neighborhood (VLSN) based heuristic solution method is described in [7]. Noting that for any fixed choice of facilities the FLPFD becomes a GAPFD, this VLSN heuristic framework uses the heuristic described in the previous section as a subroutine for a given set of open facilities. The heuristic assesses the value of various facility *open*, *close*, and *swap* moves. An open move chooses to open a previously closed facility, while a close move closes an existing open facility and a swap move takes the demands assigned to an open facility and reassigns them to a currently closed facility.

A special case of the FLPFD formulated above was defined for a problem arising in the steel manufacturing industry in [1]. For this special case $\pi_{ij} = 0$ and $a_{ij} = 0$ for all $i \in I$ and $j \in J$, and $r_{ij} = r_j$ for every $i \in I$ and for all $j \in J$, i.e., the revenue associated with a demand is independent of the resource to which it is assigned. In addition, the upper and lower bounds on the fulfillment quantity for each demand do not depend on the resource assignment, i.e., $l_{ij} = l_j$ and $u_{ij} = u_j$ for every $i \in I$ and for all $j \in J$. Strong valid inequalities are provided in [1] for tightening the LP relaxation upper bound for this special case, and a set of bin-packing-based and LP rounding heuristics are provided for obtaining fast solutions. In addition, a Lagrangian relaxation method is developed for providing good upper bounds on the problem's solution. This Lagrangian relaxation approach results in an interesting class of subproblems that also arise in the application of an exact branch-and-price approach to both the GAPFD and the FLPFD. This exact branch-and-price approach is presented in [5] and is discussed in the following chapter on decomposition approaches for location and assignment problems.

References

1. Balakrishnan A, Geunes J (2003) Production Planning with Flexible Product Specifications: An Application to Specialty Steel Manufacturing. Operations Research 51(1):94–112
2. Ingargiola G, Korsh J (1975) An Algorithm for the Solution of 0-1 Loading Problems. Operations Research 23(6):1110–1119
3. Merzifonluoğlu Y, Geunes J (2004) Requirements Planning with Order Selection and Demand Timing Flexibility. 2004 Industrial Engineering Research Conference (IERC) Proceedings
4. Merzifonluoğlu Y, Geunes J (2006) Uncapacitated Production and Location Planning Models with Demand Fulfillment Flexibility. International Journal of Production Economics 102:199–216
5. Rainwater C, Geunes J, Romeijn H (2008) Capacitated Facility Location with Single-Source Constraints and Flexible Demand. Technical report, Department of Industrial and Systems Engineering, University of Florida, Gainesville, Florida
6. Rainwater C, Geunes J, Romeijn H (2009) The Generalized Assignment Problem with Flexible Jobs. Discrete Applied Mathematics 157:49–67
7. Rainwater C, Geunes J, Romeijn H (2011) A Facility Neighborhood Search Heuristic for Capacitated Facility Location with Single-Source Constraints and Flexible Demand. Journal of Heuristics (forthcoming)

Chapter 8
Branch-and-Price Decomposition for Assignment and Location Problems

Abstract This chapter addresses a number of models with an assignment-based structure that require allocating a set of demands to a set of resources. While we have already considered several types of assignment problem in previous chapters, each of these previous models assumed that a specific assignment cost could be specified *a priori*, which depends only on the particular demand and the resource to which it is assigned. In contrast, for the models considered in this chapter, the costs associated with a resource will sometimes depend on the collective set of demands assigned to the resource. The primary approach applied for solving these problems will be a branch-and-price decomposition method. Of particular interest in applying this approach are the so-called pricing problems that arise in the decomposition. As we will see, these pricing problems will be consistent with several of the models defined and analyzed in previous chapters.

8.1 Branch-and-Price Approach

This section provides a framework for applying the branch-and-price method for the decomposition of problems containing an assignment structure. This method uses a decomposition approach for solving a problem's LP relaxation at each node in a branch-and-bound tree. Branch-and-price has been a very effective approach for solving large-scale problems with a GAP structure, particularly when the ratio of the number of demands to the number of resources is small (see [5]). Moreover, when the net profit or cost associated with a resource is a nonlinear function of the collective set of demands assigned to the resource, branch-and-price becomes an attractive alternative for solving mixed integer nonlinear optimization problems with an embedded assignment problem.

The branch-and-price method typically begins with a set partitioning formulation of the problem of interest. In particular, we wish to partition the set J of demands into m disjoint subsets J_i, $i = 1, \ldots, m$ such that $J_i \cap J_k = \emptyset$ for all $i, k \in J$ and $\bigcup_{i=1}^{m} J_i = J$, where J_i denotes the set of demands assigned to resource i. Let K_i denote the set of all possible subsets of customers that can be feasibly assigned to resource i (note that $|K_i|$ is exponential in n). Subset K_i is characterized by an n-dimensional binary vector x_i^k whose jth element equals one if customer j is included in the subset, and zero otherwise. We define λ_i^k as a binary variable equal

J. Geunes, *Demand Flexibility in Supply Chain Planning*, SpringerBriefs in Optimization, DOI 10.1007/978-1-4419-9347-2_8, © Joseph Geunes 2012

to one if subset $k \in K_i$ is selected for resource i, $i = 1, \dots, m$, and zero otherwise. Defining $\alpha_i(x_i^k)$ as the maximum net profit associated with resource i when the subset $k \in K_i$ is assigned to resource i, we formulate the set partitioning (SP) problem as follows:

$$[\text{SP}] \quad \text{Maximize} \quad \sum_{i=1}^{m} \sum_{k=1}^{K_i} \alpha_i(x_i^k)\lambda_i^k \tag{8.1}$$

$$\text{Subject to} \quad \sum_{i=1}^{m} \sum_{k=1}^{K_i} x_{ij}^k \lambda_i^k = 1, \quad j = 1, \dots, n, \tag{8.2}$$

$$\sum_{k=1}^{K_i} \lambda_i^k = 1, \quad i = 1, \dots, m, \tag{8.3}$$

$$\lambda_i^k \in \{0, 1\}, \quad i = 1, \dots, m, k = 1, \dots, K_i. \tag{8.4}$$

The objective function (8.1) maximizes the net profit from the assignment of demands to resources. The first constraint set (8.2) ensures that each demand is assigned to exactly one resource, while the second constraint set (8.3) requires choosing exactly one subset for each resource. Observe that if we permit assigning no demands to a resource (with resulting net profit of zero), then the above formulation SP allows selecting at most m subsets of demands. This is equivalent to replacing the equality in (8.3) with a less-than-or-equal-to ($\leq$) sign, which then implies that the associated dual multipliers for these constraints (when solving the LP relaxation of SP) must be nonnegative.

We have made the implicit assumption that we can enumerate all K_i subsets for every $i \in I$ and that we can evaluate the associated value of $\alpha_i(x_i^k)$, although this is not typically possible for practical problem instances. Therefore, we typically implement the solution of SP by solving its LP relaxation using column generation, where each subset K_i corresponds to a column. To do this, we first consider the LP relaxation of SP, and suppose that we have some subset of columns that ensures that a feasible solution exists. The solution of this restricted LP relaxation (containing only a subset of the columns) provides a lower bound on the optimal LP relaxation solution value. To determine whether or not the current solution is optimal for the LP relaxation, we need to determine whether or not a column exists with an attractive reduced cost that is not currently included in the restricted LP relaxation formulation. Doing this requires solving the so-called *pricing problem*.

Given an LP relaxation solution for the restricted problem, let δ_j, $j = 1, \dots, n$, and γ_i, $i = 1, \dots, m$ denote the corresponding dual variable values associated with constraints (8.2) and (8.3), respectively (recall that we replace the equality in (8.3) with a $\leq$ sign, which implies the nonnegativity of the γ_i variables). Then the reduced cost associated with a subset (column) K_i can be written as

$$\alpha_i(x_i^k) + \sum_{j \in J} x_{ij}^k \delta_j - \gamma_i. \tag{8.5}$$

Note that the sign in front of the second term above may be positive or negative, since the δ_j variables are unrestricted in sign. If (8.5) is non-positive for all $k \in K_i$ and for all $i \in I$, then no additional attractive columns exist, and the current LP relaxation solution is optimal. On the other hand, if we find some subset $k \in K_i$ for some $i \in I$ such that (8.5) is positive, then we have identified a new attractive column that must be added to the current (incomplete) formulation, and the LP relaxation must then be re-solved with this new column, and the process repeated. This iterative process repeats a finite number of times until the LP relaxation is solved.

To determine whether an attractive column exists for a resource $i \in I$, we now treat the x_i^k vector as a vector of decision variables, and maximize (8.5) over all feasible assignments of customers to the resource. In doing so, we suppress the dependence of these variables on the subset (k), and solve the following pricing problem for each resource $i \in I$:

$$[\text{PP}(i)] \quad \text{Maximize} \quad \alpha_i(x_i) + \sum_{j \in J} \delta_j x_{ij} \tag{8.6}$$

$$\text{Subject to} \quad x \in \mathscr{X}_i. \tag{8.7}$$

The set $\mathscr{X}_i$ contains all feasible demand assignments to resource i, for each $i \in I$, while the function $\alpha_i(x_i)$ determines the maximum net profit generated from this assignment. The difficulty of the pricing problem therefore depends on the form of the $\alpha_i(x_i)$ functions and the structure of the sets $\mathscr{X}_i$, and these are both highly problem dependent. The remaining sections of this chapter consider specific problem classes and the functional forms and set definitions that result for these classes.

Observe that if an optimal solution exists for PP(i) with objective function value greater than γ_i, then according to (8.5), the solution of PP(i) identifies a new attractive column for the LP relaxation, and we must re-solve the LP relaxation with this column included (in fact, we have identified a column with the highest reduced cost). If no such column exists among all resources $i \in I$, then the current LP relaxation solution is optimal. Note that identifying an attractive column only requires finding a solution to PP(i) with objective function value greater than γ_i. Once such a solution is identified, we may add the column to the LP relaxation of SP, i.e., we need not always solve PP(i) to optimality in order to identify a new column with an attractive reduced cost value. However, ultimately finding an optimal solution to the LP relaxation of SP requires solving PP(i) to optimality for all $i \in I$, in order to ensure that no additional attractive columns exist, i.e., that no non-basic variable with a positive reduced cost exists.

This decomposition method for solving the LP relaxation therefore constitutes the "price" part of the branch-and-price method. The "branch" part of the branch-and-price method results from implementing a branch-and-bound tree based on the LP relaxation solution. That is, the decomposition approach we have discussed so far leads to the solution of the LP relaxation at the so-called *root node*, as we have not explicitly required any of the λ_i^k variable to be strictly zero or one. The LP relaxation solution that results may indeed contain fractional λ_i^k variable values. When this is the case, we must branch on some variable in order to force the variable to take a binary value. Unfortunately, if we were to branch on a fractional λ_i^k variable, this would create the need to solve an LP relaxation that is structurally different from the relaxation with which we started. In other words, if we were to add a constraint of the form $\lambda_i^k \leq 0$ or $\lambda_i^k \geq 1$ to the LP relaxation of SP, then this would change the form of each of our pricing problems PP(i). As an alternative, we may omit the column corresponding to the kth subset for resource i from the formulation, which is equivalent to forcing $\lambda_i^k = 0$. Similarly, when considering the branch in which $\lambda_i^k \geq 1$, we include the kth column for resource i in the formulation, but no other columns corresponding to subsets where demand j is assigned to a resource, which forces λ_i^k to one. We can enforce these conditions in the definition of the starting set of feasible columns for the LP relaxation at a node, as well as through the definition of each set $\mathscr{X}_i$ when solving the pricing problems associated with this LP relaxation. For more details on specific implementations of the branch-and-bound scheme, please see, e.g., [3] or [5]. The remainder of this chapter discusses the form of the pricing problem that results within specific supply chain planning problems containing an underlying GAP and/or FLP structure.

8.2 Branch-and-Price for Supply Chain Planning Problems

The remainder of this chapter discusses four specific single-item supply chain planning problems that may be cast in the form of SP. The first three of these problems require assigning customer demands to a network of supply resources with single-sourcing constraints, i.e., each customer's demand must be assigned fully to one facility. As we will see, these problems produce pricing subproblems that are equivalent to particular demand selection problems already considered in earlier chapters. The fourth such problem, discussed in Sect. 8.2.4, corresponds to the classes of GAPFD and FLPFD problems defined in the previous chapter. The resulting pricing subproblems for this problem class lead to an interesting class of knapsack problems with flexible demands.

8.2.1 The Continuous-Time Single-Sourcing Problem

We first consider a set J of demands such that demand j has a constant and deterministic demand rate D_j that occurs continuously in time. Each of these demands must be assigned to some facility $i \in I$, which must satisfy all assigned demands as they occur. Resource i must manage stock in order to satisfy all assigned demands, and each resource faces costs consistent with those of the EOQ model discussed in Chap. 1. In particular, resource i incurs a holding cost of H_i per unit per unit time, and incurs a fixed order cost of S_i each time it replenishes its inventory. In addition, C_{ij} corresponds to the cost associated with satisfying customer j demand using resource i (this cost term may account for variable production and transportation costs). If x_{ij} equals one when demand j is assigned to resource i, then resource i faces an EOQ problem with demand rate $\sum_{j\in J} D_j x_{ij}$. The goal of this continuous-time single-sourcing problem is then to determine the assignment of demands to resources that minimizes cost across a network of resources, each of which fulfills its assigned demands.

This problem class might arise in a number of particular contexts. For example, if each resource corresponds to a distribution center that fulfills market demands as they arise (e.g., using a package shipping service), then this problem requires assigning markets to distribution centers. We are interested in characterizing the form of the function $\alpha_i(x_i)$ and the set $\mathscr{X}_i$, as well as the structure of the pricing problem PP(i), for each $i \in I$, which will permit application of the branch-and-price method described in the previous section. For a detailed discussion of this problem class, several problem variants, and computational test results using the branch-and-price approach, please see [2].

Because each facility faces an EOQ-type problem and must meet all assigned demand, the function $\alpha_i(x_i)$ corresponds to the negative of the inventory ordering and holding costs at facility i. In particular,

$$\alpha_i(x_i) = -\left\{\sum_{j\in J} C_{ij}x_{ij} + \sqrt{2S_i H_i \sum_{j\in J} D_j x_{ij}}\right\}. \tag{8.8}$$

The set $\mathscr{X}_i$ defines the allowable assignments to facility i; in the absence of any specific capacity or budget constraint we have $\mathscr{X}_i = \{0,1\}^n$. Given the definition of $\alpha_i(x_i)$ above, the pricing problem is equivalent to the following problem:

$$[\text{PP}_{\text{EOQ}}(i)] \quad \text{Maximize} \quad \sum_{j\in J}(\delta_j - C_{ij})x_{ij} - \sqrt{2S_i H_i \sum_{j\in J} D_j x_{ij}} \tag{8.9}$$

$$\text{Subject to} \quad x_{ij} \in \{0,1\}, \quad j = 1,\ldots,n. \tag{8.10}$$

This pricing problem is equivalent to the EOQ problem with market choice discussed in Sect. 3.1. As a result, this pricing problem can be solved in polynomial time for given values of the dual multipliers δ_j for all $j \in J$. When resource production or inventory capacities exist, then the set $\mathscr{X}_i$ must be redefined to account for these constraints. For details on handling such constraints, as well as heuristic solution approaches for the continuous-time single-sourcing problem, please see [2].

8.2.2 Single-Period Demand Allocation with Uncertainty

This section considers the assignment of a collection of demands to multiple resources in a single-period setting with uncertain demands. That is, we consider a set J of n independent demands for a single sales period, where demand $j \in J$ in the period is a normally distributed random variable with expected value μ_j and standard deviation σ_j. Each demand must be assigned to some resource $i \in I$, where I

denotes a set of m resources. Letting D_j denote the random variable for demand j, then if x_{ij} equals one when demand j is assigned to resource i (and zero otherwise), the demand observed by resource i equals $\sum_{j\in J} D_j x_{ij}$. Under binary x_{ij} variables and the independence and normality of demands, the expected value of the demand observed by resource i equals $\sum_{j\in J} \mu_j x_{ij}$, while the variance equals $\sum_{j\in J} \sigma_j^2 x_{ij}$.

In addition to determining the assignment of demands to resources, we also wish to determine each resource's capacity level Q_i for meeting assigned demands. Suppose that the cost per unit produced by resource i equals C_i, while resource i incurs a unit understock cost of B_i for each demand in excess of Q_i, while incurring a unit overstock cost of H_i for leftover units after assigned demand is realized (a negative value of H_i corresponds to a salvage value, and we assume $B_i > C_i > -H_i$). We assume that B_i corresponds to a unit emergency supply cost and that demands in excess of supply are satisfied through an emergency shipment.

The optimal resource i capacity level depends on the assignment vector x_i. In particular, if $F_{x_i}(\cdot)$ denotes the CDF of assigned resource i demands, then the optimal capacity level given the n-dimensional assignment vector x_i is determined by the equation

$$F_{x_i}\left(Q_i^*(x_i)\right) = \frac{B_i - C_i}{B_i + H_i}. \tag{8.11}$$

Letting z_i^* denote the standard normal value such that the CDF of the standard unit normal distribution equals the right-hand side of (8.11), then we can write $Q_i^*(x_i)$ explicitly as

$$Q_i^*(x_i) = \sum_{j\in J} \mu_j x_{ij} + z_i^* \sqrt{\sum_{j\in J} \sigma_j^2 x_{ij}}. \tag{8.12}$$

Using the approaches described in Sect. 1.2.2 and in [4], if r_{ij} denotes the unit revenue obtained when serving demand j using resource i, then the expected profit associated with facility i can be written as

$$\alpha_i(x_i) = \sum_{j\in J} \hat{r}_{ij} x_{ij} - K_i\left(z_i^*\right) \sqrt{\sum_{j\in J} \sigma_j^2 x_{ij}}, \tag{8.13}$$

where $\hat{r}_{ij} = (r_{ij} - C_i + H_i)\mu_j$ and $K_i(z_i^*) = (C_i + H_i)z_i^* + (B_i + H_i)L(z_i^*)$ (note that we can also easily account for a fixed cost S_{ij} for assigning demand j to resource i by simply subtracting S_{ij} from $\hat{r}_{ij}$). Then the pricing problem under the branch-and-price approach to solving this assignment problem, assuming no explicit capacity constraints on the Q_i values, is stated as follows:

$$[\text{PP}_{\text{NV}}(i)] \quad \text{Maximize} \quad \sum_{j\in J} (\hat{r}_{ij} + \delta_j) x_{ij} - K_i\left(z_i^*\right) \sqrt{\sum_{j\in J} \sigma_j^2 x_{ij}} \tag{8.14}$$

$$\text{Subject to} \quad x_{ij} \in \{0, 1\}, \quad j = 1, \ldots, n. \tag{8.15}$$

The pricing problem $\text{PP}_{\text{NV}}(i)$ above is equivalent to the selective newsvendor problem (SNP) defined in Sect. 4.2, where we showed that this problem can be easily solved in polynomial time. The branch-and-price approach to this problem class is discussed in greater detail in [4], as are methods for solving the problem under capacity constraints on the Q_i values and heuristic solution methods for quickly finding feasible solutions.

8.2.3 Integrated Facility Location and Production Planning

We next describe a model for facility location planning that integrates time-phased production allocation decisions with facility location decisions. The traditional facility location problem (FLP), defined in Sect. 1.2.6, does not contain a time dimension. That is, a set J of n demands exists such that D_j denotes the quantity associated with demand j for $j \in J$, and the FLP is thus effectively a single-period problem. The FLP uses an assignment cost of c_{ij} to capture the cost associated with using facility i to satisfy demand j. However, this assignment cost is typically a rough-cut approximation that attempts to capture the production cost at facility i as well as the cost to deliver the output to the customer. In order to more accurately capture production costs, we consider a model that accounts for fixed and variable production costs as well as inventory holding costs at the facilities directly within the location planning problem.

More specifically, we consider a time horizon containing T periods, such that n_t demands exist in period t, for $t = 1, \ldots, T$, and we let d_{jt} denote the quantity associated with demand j in period t. We wish to assign these demands to a set of interchangeable supply facilities selected from among a set I of potential locations. A fixed cost $\mathscr{F}_i$ is incurred when operating facility i over the time horizon for each $i \in I$. The cost $\mathscr{F}_i$ may be an amortized (e.g., yearly) cost, while the demands may correspond to a set of demands that are expected to repeat from year to year.

Each facility follows a production plan in order to meet the assigned demands. Any positive production quantity in period t at facility i results in a fixed production cost of S_{it}, as well as a unit cost of C_{it} for each unit produced and a holding cost of H_{it} for inventory remaining at facility i at the end of period t. For simplicity we assume that the facilities will not be capacity constrained with respect to production planning (the model we define is a special case of that defined in [6]). We wish to simultaneously determine the set of open facilities, the assignment of demands to facilities, and the minimum production and inventory costs at the facilities for the given demand assignments. We therefore define y_i^f as a binary variable equal to one if facility i is open (and zero otherwise) for all $i \in I$. We also define y_{it}^p as a binary variable equal to one if production occurs at facility i in period t (and zero otherwise) for all $i \in I$ and $t = 1, \ldots, T$. We next define x_{ijt} as a binary variable equal to one if the jth demand in period t is assigned to facility i, and let Q_{it} and I_{it} denote the production and inventory quantities at facility i in period t, respectively. This integrated location and production planning problem (ILPP) can then be formulated as follows:

$$\text{[ILPP]} \quad \text{Minimize} \quad \sum_{i \in I} \left\{ \mathscr{F}_i y_i^f + P_i(x_{i\cdot\cdot}) \right\} \tag{8.16}$$

$$\text{Subject to} \quad \sum_{i \in I} x_{ijt} \geq 1, \quad t = 1, \ldots, T, j = 1, \ldots, n_t, \tag{8.17}$$

$$x_{ijt} \leq y_i^f, \quad i \in I, t = 1, \ldots, T, j = 1, \ldots, n_t, \tag{8.18}$$

$$x_{ijt} \in \{0, 1\}, \quad i \in I, t = 1, \ldots, T, j = 1, \tag{8.19}$$

$$y_i^f \in \{0, 1\}, \quad i \in I. \tag{8.20}$$

The objective function (8.16) contains the function $P_i(x_{i\cdot\cdot})$ which computes the minimum production and inventory costs at facility i, given the assignments implied by the vector $x_{i\cdot\cdot}$, which is a $t \times n_t$ vector such that element (t, n_t) equals one if the jth demand in period t is assigned to facility i (and zero otherwise). The first constraint set (8.17) requires assigning the jth demand in period t to some facility, while the second constraint set (8.18) disallows assignments to any facility that is not open. Computing $P_i(x_{i\cdot\cdot})$ requires solving an instance of the ELSP, where the demand in period t at facility i equals $\sum_{j=1}^{n_t} d_{jt} x_{ijt}$.

We can formulate this problem as a set partitioning problem in which the resources correspond to facilities. In order to translate the ILPP formulation to that of SP, we first recognize that each (j, t) pair in the ILPP formulation corresponds to a unique demand index j in the set partitioning formulation, where the set J now consists of all (j, t) pairs. The set partitioning formulation requires defining the function $\alpha_i(x_i)$, which in this setting corresponds to the negative of the minimum cost associated with facility i, given the assignment vector x_i (note that we can translate a given vector $x_{i\cdot\cdot}$ in ILPP to a vector x_i in SP in the same way we translate (j, t) pairs to a single index j). In particular, assuming $x_{i\cdot\cdot} \neq 0$, then for each $i \in I$ we have

$$\alpha_i(x_i) = -\left\{ \mathscr{F}_i + P_i(x_{i\cdot\cdot}) \right\}. \tag{8.21}$$

Solving the pricing problem for facility i is then equivalent to solving the following problem:

$$\text{Maximize} \quad \sum_{t=1}^{T} \left\{ \sum_{j=1}^{n_t} \delta_{jt} x_{ijt} - P_i(x_{i\cdot\cdot}) \right\} - \mathscr{F}_i \tag{8.22}$$

$$\text{Subject to} \quad x_{ijt} \in \{0, 1\}, \quad t = 1, \ldots, T, j = 1, \ldots, n_t. \tag{8.23}$$

Recall that determining the value of $P_i(x_{i\cdot\cdot})$ for any given values of the x_{ijt} variables requires solving an instance of the ELSP. As a result, the above pricing problem is equivalent to solving the following problem:

$$[\text{PP}_{\text{DSP}}(i)] \quad \text{Maximize} \quad \sum_{t=1}^{T}\left\{\sum_{j=1}^{n_t}\delta_{jt}x_{ijt} - S_{it}y_{it}^{p} - C_{it}Q_{it} - H_{it}I_{it}\right\} - \mathscr{F}_i \tag{8.24}$$

$$\text{Subject to} \quad I_{it} = Q_{it} + I_{i,t-1} - \sum_{j=1}^{n_t} d_{jt}x_{ijt}, \quad t = 1, \ldots, T, \tag{8.25}$$

$$Q_{it} \leq M_t y_{it}^{p}, \quad t = 1, \ldots, T, \tag{8.26}$$

$$I_{i0} = 0, \qquad Q_{it}, I_{it} \geq 0, \quad t = 1, \ldots, T, \tag{8.27}$$

$$0 \leq x_{ijt} \leq 1, \quad t = 1, \ldots, T, \; j = 1, \ldots, n_t, \tag{8.28}$$

$$y_{it}^{p} \in \{0, 1\}, \quad t = 1, \ldots, T. \tag{8.29}$$

The pricing problem $\text{PP}_{\text{DSP}}(i)$ above is equivalent to the demand selection problem (DSP) defined in Sect. 5.1, which can be solved in polynomial time. Computational experience with the branch-and-price approach we have outlined for the ILPP problem is presented in [6], along with a discussion of more general production cost functions, and an analysis of the impacts of forecasting errors on the model's performance.

8.2.4 The GAP and FLP with Flexible Demands

Chapter 7 defined models for generalized assignment and facility location problems with demand specification flexibility, the so-called GAPFD and FLPFD. The crucial difference between these two problems lies in the inclusion of a fixed cost for opening a facility in the latter problem, while the GAPFD corresponds to the special case of the FLPFD with zero fixed costs for facilities. Because of this, the branch-and-price approach is virtually the same for both of these models. Both models require assigning demands to resources, as with each of the models we have discussed thus far in this chapter. They can therefore be cast as set partitioning (SP) problems where the function $\alpha_i(x_i)$ corresponds to the maximum profit associated with resource i for the given assignment of demands to the resource implied by the vector x_i. As a result, we will focus on the pricing problem that must be solved in the branch-and-price approach.

Our analyses in the previous three subsections led to a definition of the pricing problem's feasible set, $\mathscr{X}_i$, for any $i \in I$, containing a very simple form, i.e., where the only constraints on the x_{ij} variables were the binary restrictions (although we alluded to problems involving additional constraints, we did not provide a detailed analysis of such problems, instead referring to relevant references). The pricing problem in the case of the GAPFD and the FLPFD requires explicit recognition of the additional resource capacity and demand bounding constraints in order to properly model the problem. We will consider the pricing problem for the FLPFD, recognizing that the pricing problem for the GAPFD corresponds to the special case with each facility fixed cost term S_i equal to zero.

Unlike our previous pricing problems, determining the value of $\alpha_i(x_i)$ requires solving an optimization problem, even for a fixed value of the vector x_i. In particular, for a given facility i, the value of $\alpha_i(x_i)$ for a fixed vector x_i is obtained by solving the following linear program in the v_{ij} variables:

$$\text{Maximize} \quad \sum_{j \in J} r_{ij} v_{ij} \tag{8.30}$$

$$\text{Subject to} \quad \sum_{j \in J} (a_{ij} x_{ij} + v_{ij}) \le b_i, \tag{8.31}$$

$$l_{ij} x_{ij} \le v_{ij} \le u_{ij} x_{ij}, \quad j \in J. \tag{8.32}$$

Let $g_i(x_i)$ denote the optimal value of the above linear program for a given vector x_i. Using the definition of the function $\alpha_i(x_i)$ in the case of the FLPFD as well as the general form of the pricing problem (8.6)–(8.7), we obtain the following pricing problem for facility i in the analysis of the FLPFD:

$$[\text{PP}_{\text{KPFD}}(i)] \quad \text{Maximize} \quad \sum_{j \in J} \left\{ (\pi_{ij} + \delta_j) x_{ij} + g_i(x_i) \right\} - S_i \tag{8.33}$$

$$\text{Subject to} \quad x \in \mathscr{X}_i. \tag{8.34}$$

In the above formulation, $\mathscr{X}_i$ is the set of vectors x_i such that (8.31)–(8.32) is nonempty. The optimal pricing problem solution can be obtained by solving the following *knapsack problem with flexible demands* (KPFD) with $\hat{\pi}_{ij} = \pi_{ij} + \delta_j$ for each $j \in J$:

$$[\text{KPFD}] \quad \text{Maximize} \quad \sum_{j \in J} \{ \hat{\pi}_{ij} x_{ij} + r_{ij} v_{ij} \} \tag{8.35}$$

$$\text{Subject to} \quad \sum_{j \in J} (a_{ij} x_{ij} + v_{ij}) \le b_i, \tag{8.36}$$

$$l_{ij} x_{ij} \le v_{ij} \le u_{ij} x_{ij}, \quad j \in J, \tag{8.37}$$

$$x_{ij} \in \{0, 1\}, \quad j \in J. \tag{8.38}$$

If the optimal solution value of the above KPFD problem is greater than $S_i + \gamma_i$, then an attractive column for the SP formulation of the FLPFD has been identified. The KPFD problem was analyzed in [1], which provides an $\mathscr{O}(U_i b_i)$ dynamic programming algorithm for problems with integer data, where $U_i = \sum_{j \in J} (u_{ij} - l_{ij} + 1)$. The pricing problem for the FLPFD (and GAPFD) can therefore be solved in pseudopolynomial time. A more general version of the KPFD is analyzed in [3], which considers general nonlinear revenue functions of the form $r_{ij}(v_{ij})$ instead of the linear functions considered above. The results of extensive computational studies of the branch-and-price approach for both the GAPFD and the FLPFD can be found in [3], under various assumptions on the structure of the revenue functions.

References

1. Balakrishnan A, Geunes J (2003) Production Planning with Flexible Product Specifications: An Application to Specialty Steel Manufacturing. Operations Research 51(1):94–112
2. Huang W, Geunes J, Romeijn H (2005) The Continuous-Time Single-Sourcing Problem with Capacity Expansion Opportunities. Naval Research Logistics 52(3):193–211
3. Rainwater C, Geunes J, Romeijn H (2008) Capacitated Facility Location with Single-Source Constraints and Flexible Demand. Technical report, Department of Industrial and Systems Engineering, University of Florida, Gainesville, Florida
4. Taaffe K, Geunes J, Romeijn H (2010) Supply Capacity Acquisition and Allocation with Uncertain Customer Demands. European Journal of Operational Research 204:263–273
5. Savelsbergh M (1997) A Branch-And-Price Algorithm for the Generalized Assignment Problem. Operations Research 45(6):831–841
6. Sharkey T, Geunes J, Romeijn H, Shen Z (2011) Exact Algorithms for Integrated Facility Location and Production Planning Problems. Naval Research Logistics 58(5):419–436

Part IV
Research Directions and Modeling Challenges

Chapter 9
Research Challenges in Supply Chain Planning with Flexible Demand

Abstract This chapter discusses the limitations of the models we have discussed throughout the book, and uses these limitations to characterize challenging future research directions. In doing so, we discuss the relation of the models we have considered to practice, and their potential for use in practical applications. We also consider the potential for expanding the definition of demand flexibility, as well as the technical difficulties inherent in meeting the challenges implied by potential future research avenues.

9.1 Dimensions of Demand Flexibility Modeling

The models presented in this book have taken a number of classical operations problems and generalized them to permit some flexibility in the requirements a supplier must meet. In other words, whereas classical operations models have typically considered demands or requirements as exogenous or "given" parameters, this book effectively views these requirements as inherent decision variables. As discussed in Chap. 2, models that consider price-dependent demand have taken this view in the literature for many years. That is, when prices are decision variables, and prices imply demand levels, this constitutes one form of demand flexibility wherein a supplier attempts to determine optimal demand levels.

The models presented in this book have taken a more direct view, treating the demand levels themselves as decision variables. This view recognizes the fact that many producers have some level of discretion in determining the customer orders they are willing to accept. We also consider practical contexts in which customers allow some degree of demand specification flexibility. Although we permitted certain degrees of demand flexibility in the models we analyzed, some may consider this view of demand flexibility to be somewhat limited.

For example, the models in this book do not permit *temporal flexibility*. That is, given a demand in some period t, we assumed that the customer is inflexible with respect to time, i.e., the demand must be met in period t or not at all. As one example of a type of temporal flexibility, we might consider the availability of backlogging (or even early delivery of) some or all of a customer's demand in a period. While many of the models we have discussed generalize fairly easily to the case of backlogging under certain backlogging cost assumptions, temporal flexibility in demand

J. Geunes, *Demand Flexibility in Supply Chain Planning*,
SpringerBriefs in Optimization,
DOI 10.1007/978-1-4419-9347-2_9, © Joseph Geunes 2012

satisfaction may provide substantial economic value to a supplier (an ELSP with demand time windows is considered in [7], while a problem with demand selection and time windows is analyzed in [9]). Because customers are often willing to permit demand fulfillment timing flexibility in exchange for a price break, incorporating such temporal flexibility into operations models has potential for practical impact.

The models discussed throughout this book have also been limited to single-product contexts. Beyond the practical value of generalizing these models to account for potential complexities introduced in multi-product settings, these multi-product settings can lead to another dimension of demand flexibility involving product substitutions. Product and component substitutability can lead to substantial economic benefits for a supplier (see, e.g., [2, 4]), but this requires an understanding of customers' tolerances in terms of accepting a substitute product or component. Integrated modeling of production planning and end product substitutions (given customers' willingness to accept substitute products and their reservation prices for substitutes) can aid in product line design decisions that maximize profitability (see [4]). Models that permit demand selection, pricing, demand specification flexibility, and product substitutions provide an interesting direction for future research on the benefits of demand flexibility to a supplier.

9.2 Model Limitations

The models we considered throughout this book have attempted to balance between practical applications and model tractability. As a result, some of the assumptions employed may be overly restrictive for certain practical settings. The majority of the analysis has focused on deterministic models, and the stochastic models we discussed certainly have limited application (recall that we only considered single-period models with demand uncertainty, although these easily generalize to infinite-horizon cases with stationary demands and variable costs, zero fixed order cost, and unlimited capacities). These models thus serve as a rough approximation of the corresponding situation faced in practice, and may therefore be used in planning phases, along with sensitivity analysis to determine how structural decisions (e.g., order acceptance, facility location) might change as model parameters vary.

The models that we considered involving demand uncertainty used an assumption of statistical independence of individual demands that are normally distributed. One interesting direction for future research would permit correlation of demands, for example, in the selective newsvendor problem (SNP) with normal demands. In this case, if ρ_{jk} denotes the correlation coefficient for demands j and k, then the SNP (originally discussed in Sect. 4.2) would be written as follows:

$$\text{Maximize}_{x\in\{0,1\}^n} \sum_{j\in J} r_j\mu_j x_j - K(z^*)\sqrt{\sum_{j\in J}\sigma_j^2 x_j + 2\sum_{j\in J}\sum_{k>j}\rho_{jk}\sigma_j\sigma_k x_j x_k}. \quad (9.1)$$

The additional covariance term under the square root function leads to nontrivial mathematical difficulties, and eliminates the ability to implement a sorting-based algorithm for determining an optimal solution.

Another simplifying assumption we made in the selective newsvendor setting was the ability to set capacity, Q, based on the optimal critical fractile value, according to Eq. (1.7). If this capacity is set to some fixed value of b, then we have a substantially more difficult problem (this problem is sometimes referred to as a static stochastic knapsack problem; see [3, 6], or [10]). Under this assumption we cannot use a simple sorting approach to solve the problem, although the problem's continuous relaxation can be formulated as a convex program. The normal demand assumption is also clearly an approximation made for tractability, and dealing with non-normal demand distributions can be challenging (note that a static stochastic knapsack problem with independent Poisson distributed demands is considered in [1]). Beyond dealing with other demand distributions, more sophisticated models are needed to handle a broader set of problems involving stochastic demands beyond those that can be expressed as single-period problems.

In dealing with generalizations of the ELSP (i.e., the demand selection and market selection problems in Chaps. 5 and 6), we have primarily shied away from dealing with finite production capacities. As we showed in Chap. 6, the market selection problem is $\mathscr{N}\mathscr{P}$-Hard even in the absence of production capacities. We did note, however, in Chap. 5, that the demand selection problem with time-invariant capacities is polynomially solvable, albeit with a high-order polynomial. Dealing with such capacities is necessary in practice, and heuristic methods for dealing with such problems are necessary for handling practical problem contexts.

9.3 Limitations of Branch-and-Price Decomposition

The branch-and-price method discussed in Chap. 8 is a very useful approach for decomposing a large-scale optimization problem into a set of subproblems, each of which is considerably easier to solve than the original problem. This approach clearly has some limitations, as it is an exact method that we have applied for solving $\mathscr{N}\mathscr{P}$-Hard problems. Thus, while branch-and-price often permits solving larger problems than would otherwise be solvable (e.g., by using a commercial solver to attack a large-scale problem formulation), we eventually run into memory or time constraints with this method as well. This is because the set partitioning problem, which is used as the basis for the branch-and-price method, is itself an $\mathscr{N}\mathscr{P}$-Hard problem in general (despite the fact that the LP relaxation of set partitioning is often very tight with respect to the optimal integer solution). Moreover, the pricing problems we have discussed are, in some cases, difficult problems themselves. For example, the pricing problems for the GAPFD and the FLPFD were solvable in pseudopolynomial time when the revenue functions were linear. Thus, if the problem data values are large, solving a single instance of the pricing problem can itself be time consuming. Under general nonlinear revenue functions, the pricing problems may themselves be $\mathscr{N}\mathscr{P}$-Hard optimization problems. Clearly then, for large enough problem instances, the branch-and-price method may not serve as a practical solution approach, and heuristic methods must be developed and applied, leaving heuristic development for large-scale problems as a promising avenue for further research.

9.4 Further Generalizations and Approximation Algorithms

One of the methods discussed in Chap. 6 for solving the $\mathcal{NP}$-Hard market selection problem (MSP) was an LP-rounding based approximation algorithm. Recall that this method begins with the solution of the problem's LP relaxation. Given a solution vector $\hat{w}$ of market selection variables for this LP relaxation, and given a fraction β, we round the market selection variable for market j to one if $\hat{w}_j \geq \beta$ and we round down to zero otherwise (recall also that rounding up is equivalent to rejecting a market, while rounding down is equivalent to selecting a market). Given the rounded market selection variables, we can then solve an instance of the ELSP with the selected markets to obtain a feasible solution for the MSP. As we noted in Chap. 6, by the appropriate selection of m values of β, we were able to show that the application of this heuristic approach m times permits obtaining a feasible solution for the MSP whose solution value is no more than 1.582 times the minimum cost solution.

In [5], a more general class of market selection problems is considered, along with a general version of this rounding-based approximation algorithm. In particular, the following general problem class, denoted by P, is considered in [5]:

$$[\text{P}] \qquad \min_{w \in \{0,1\}^n} rw - \phi(w). \tag{9.2}$$

In problem P, r is an n-vector of market revenue values, w is an n-vector of market selection variables, and $\phi(w)$ is a nonnegative (cost) function that depends on w. For example, in the case of the MSP, $\phi(w)$ is the minimum cost ELSP solution for a given choice w of market selection variables.

Evaluating $\phi(w)$ requires solving an optimization problem in which w serves as a set of parameters. Let $\Phi(w)$ denote this corresponding optimization problem. Suppose that a lower bounding function $\bar{\phi}(w)$ exists for $\phi(w)$ (with $\bar{\phi}(w) \leq \phi(w)$ for all $w \in [0,1]^n$), and that it is possible to efficiently find a feasible solution to $\Phi(w)$ with solution value no more than $\alpha\bar{\phi}(w)$ for some $\alpha \geq 1$. Because $\bar{\phi}(w)$ is a lower bounding function, and because we can find a feasible solution to $\Phi(w)$ with value no higher than $\alpha\bar{\phi}(w)$, this implies that we have an α-approximation algorithm for $\Phi(w)$ (in the case of the MSP, $\Phi(w)$ is an instance of the ELSP, $\bar{\phi}(w)$ is the LP relaxation solution, and we can quickly find a feasible solution to $\Phi(w)$ with value no more than $\bar{\phi}(w)$ because the LP relaxation is tight; this implies that we have a 1-approximation algorithm for $\Phi(w)$ in this case).

Suppose further that the continuous relaxation of P can be solved efficiently when $\phi(w)$ is replaced by $\bar{\phi}(w)$ (where $w \in \{0,1\}^n$ is also replaced by $w \in [0,1]^n$).

Next, given any vector $w \in [0,1]^n$ and a fraction β with $0 \le \beta < 1$, let $[w]_\beta$ denote the rounded vector of w_j values (using our rounding rules), and assume that $\bar{\phi}([w]_\beta) \le \frac{1}{1-\beta}\bar{\phi}(w)$. This last condition establishes an upper bound on the minimum cost solution to $\Phi(w)$ when using the rounded solution vector.

The approximation algorithm provided in [5] works as follows. We first solve the continuous relaxation of problem P when $\phi(w)$ is replaced with the lower bounding function $\bar{\phi}(w)$. Let $\tilde{w}$ denote the corresponding (fractional) market selection vector solution. Next, given β, we round $\tilde{w}$ to obtain $[\tilde{w}]_\beta$, and then apply the α-approximation algorithm to obtain a solution to $\Phi([\tilde{w}]_\beta)$, and a corresponding feasible solution to problem P.

Observe that $r[\tilde{w}]_\beta \le (1/\beta)r\tilde{w}$ and we obtain a solution using the approximation algorithm with cost no greater than $\alpha\bar{\phi}([\tilde{w}]_\beta)$. Because $\bar{\phi}([\tilde{w}]_\beta) \le \frac{1}{1-\beta}\bar{\phi}(\tilde{w})$, this implies that our algorithm gives a feasible solution to P with cost no greater than

$$\frac{1}{\beta}r\tilde{w} + \frac{\alpha}{1-\beta}\bar{\phi}(\tilde{w}) \le \max\left\{\frac{1}{\beta}, \frac{\alpha}{1-\beta}\right\}\left(r\tilde{w} + \bar{\phi}(\tilde{w})\right) \le \max\left\{\frac{1}{\beta}, \frac{\alpha}{1-\beta}\right\}Z^*,$$

where Z^* is the optimal solution value for problem P, and, $r\tilde{w} + \bar{\phi}(\tilde{w})$, the optimal relaxation solution value, provides a lower bound on Z^*. Setting $\beta = 1/(\alpha+1)$ gives $1/\beta = \alpha/(1-\beta) = \alpha + 1$. Thus, when $\beta = 1/\alpha + 1$, this approximation algorithm provides a feasible solution with a performance guarantee of $\alpha + 1$ (see [5]).

In our approximation algorithm in Chap. 6 for the MSP, because $\alpha = 1$, this implies a worst-case performance guarantee of 2 for the MSP. As discussed in Chap. 6, by randomizing the algorithm, and then derandomizing, this worst-case guarantee was improved to 1.582. Similarly, for the general problem class P, it is possible to show that if an α-approximation algorithm exists for problem $\Phi(w)$, then we can use the rounding approach and the α-approximation algorithm to obtain a solution to problem P with a worst-case performance guarantee of $\frac{1}{1-e^{-1/\alpha}}$ (see [5]).

Because problem P is posed quite generally, this result has broad implications for generalizing several $\mathcal{NP}$-Hard problems to permit demand selection, as noted in [5]. We illustrate this using the so-called one-warehouse multi-retailer (OWMR) problem as an example. This problem contains a set of n retailers that are supplied with inventory from a centralized warehouse, which replenishes its supply using an external supplier. Each retailer faces deterministic demands in every period over a finite time horizon of length T. A fixed cost is incurred when any retailer places a replenishment order with the warehouse, and when the warehouse places a replenishment order with its supplier. The objective of the OWMR problem is to minimize total system-wide fixed ordering plus inventory holding costs.

In the market-selection version of the OWMR problem, retailer i has a demand of d^i_{jt} from market j in period t, for $j \in J$, $t = 1, \ldots, T$, and $i = 1, \ldots, m$, where J denotes a set of markets. The goal is to determine a selection of markets to serve, which then determines the demands that the retailers must face. Thus, for a given selection of markets, the problem reduces to the OWMR problem. Because the joint replenishment problem (JRP) is the special case of the OWMR problem in which the warehouse inventory cost is infinite (and the warehouse, therefore, holds no inventory), all of the results that hold for the OWMR problem with market selection also hold for the JRP with market selection. As shown in [8], a 1.8-approximation algorithm exists for the OWMR. Based on the previous results we have discussed for problem P, this implies a

$$\frac{1}{1 - e^{-1/1.8}} \approx 2.35$$

approximation algorithm for the market selection version of the OWMR problem and, therefore, the JRP as well. Additional problem generalizations and associated worst-case performance bounds are discussed in [5], including problems with soft-capacities, problems with uncertain market revenues and demands, an assembly problem, and specific types of facility location problems with market choice.

The preceding discussion provides an indication of the potential that exists for additional research involving selection problems, particularly in terms of identifying general models and results that may apply broadly to problem classes that permit demand flexibility. While these results correspond to approximation algorithms with worst-case performance guarantees, analysis of exact and heuristic approaches for solving general problem classes with demand flexibility also serves as a promising direction for continued research.

References

1. Ağralı S, Geunes J (2009) A Single-Resource Allocation Problem with Poisson Resource Requirements. Optimization Letters 3:559–571
2. Balakrishnan A, Geunes J (2000) Requirements Planning with Substitutions: Exploiting Bill-of-Materials Flexibility in Production Planning. Manufacturing & Service Operations Management 2(2):166–185
3. Barnhart C, Cohn A (1998) The Stochastic Knapsack Problem with Random Weights: A Heuristic Approach to Robust Transportation Planning. In: Proceedings of Tristan III, Puerto Rico, 17–23
4. Ervolina T, Ettl M, Lee Y, Peters D (2009) Managing Product Availability in an Assemble-to-Order Supply Chain with Multiple Customer Segments. OR Spectrum 31:257–280
5. Geunes J, Levi R, Romeijn H, Shmoys D (2011) Approximation Algorithms for Supply Chain Planning and Logistics Problems with Market Choice. Mathematical Programming, Series A 130(1):85–106
6. Kleywegt A, Shapiro A, Homem-de-Mello T (2001) The sample average approximation method for stochastic discrete optimization. SIAM Journal of Optimization 12(2):479–502
7. Lee C, Çetinkaya S, Wagelmans A (2001) A Dynamic Lot Sizing Model with Demand Time Windows. Management Science 47(10):1384–1395
8. Levi R, Roundy R, Shmoys D, Sviridenko M (2008) A Constant Approximation Algorithm for the One-Warehouse Multi-Retailer Problem. Management Science 54:763–776
9. Merzifonluoğlu Y, Geunes J (2006) Uncapacitated Production and Location Planning Models with Demand Fulfillment Flexibility. International Journal of Production Economics 102:199–216
10. Merzifonluoğlu Y, Geunes J, Romeijn H (2011) The Static Stochastic Knapsack Problem with Normally Distributed Item Sizes. Mathematical Programming, Series A (forthcoming)

CPSIA information can be obtained at www.ICGtesting.com
Printed in the USA
LVOW10s2204110314

377030LV00002B/23/P

9 781441 993465